IMPROVING LABORATORY ANIMAL WELFARE
A CRITICAL EVALUATION OF THE QUALITY OF LITERATURE SEARCH STRATEGIES USED IN SYSTEMATIC REVIEWS OF ANIMAL EXPERIMENTS

by

Nieky van Veggel

MSc Animal Biology and Welfare

September 2010

Submitted as part requirement for the Degree of MSc Animal Biology and Welfare, Writtle College and University of Essex

Abstract

Systematic reviews of animal experiments (SR of AE) have a great potential to improve laboratory animal welfare. However, at the moment they are not commonly used in laboratory animal science, even though they are generally regarded by professionals in the field of evidence-based medicine as the highest level of medical evidence, and they are already standard practice for clinical studies. Systematic reviews of animal experiments allow others to replicate and build on previously published work, diminish the number of animals needed in animal experimentation, improve animal welfare and improve patient safety.

Collecting and analysing all available literature before starting an animal experiment is important and it is indispensable when writing a systematic review of animal research. Writing such reviews prevents unnecessary duplication of animal studies and thus unnecessary animal use. One of the factors currently inhibiting the production of high-quality SRs in laboratory animal science is the fact that searching for all available literature concerning animal experiments is difficult. Therefore, this study set out to assess the quality of the search strategies used in current systematic reviews of animal experiments, in order to evaluate the areas that need improving.

An assessment system was created based on several guidelines for human health care systematic reviews. The literature search strategy used was aimed at PubMed and EMBASE, the two most frequently used bibliographic databases in biomedical science, and yielded 80 systematic reviews of animal experiments after the study selection phase.

The results show an increase in the number of SR of AE between 2005 and 2010. Also, an increase in overall quality of the search strategies used in current SR of AE was found between 2005 and 2010 and especially between 2008 and 2010. The quality of the individual assessed items showed great differences, with the best item scoring 98.1% and the worst scoring item 13.8%. References by journal publishers to SR guidelines appeared to increase the quality of SR of AE. Furthermore, this study showed differences in quality exist between human medicine systematic reviews and SR of AE.

In conclusion, the quality of systematic reviews of animal experiments has increased over the years, but is far from perfect. Guidelines for writing systematic reviews seem a useful tool, since reference to systematic review guidelines appeared to increase quality. Futher research should aim at developing guidelines specifically for systematic reviews for animal experiments, in order to make them more feasible and more standardised. Systematic reviews of animal experiments have the potential to increase laboratory animal welfare, but only if they become of consistent, high, quality and fully accepted in the field of laboratory animal science.

Contents

Abstract ii

Contents iv

Abbreviations vi

List of Figures vii

List of Tables viii

List of Appendices viii

Acknowledgements ix

1 Introduction 1

2 Literature Review 4
   2.1 Laboratory animal welfare 4
   2.2 Systematic reviews 5
   2.3 Designing a bibliographic search strategy 7
   2.4 Reporting a search strategy 9
   2.5 The influence of a literature search strategy on systematic reviews and laboratory animal welfare 10

3 Materials and Methods 12
   3.1 Search strategy for identification of studies 12
   3.2 Study selection 12
   3.3 Quality assessment 13
   3.4 Data extraction 13
   3.5 Statistical analysis 13
   3.6 Developing a step-by-step search and reporting guide 14

4 Results 16
   4.1 Literature search 16
   4.2 Study selection 17
   4.3 Publications per year 17
   4.4 Assessment items 19
   4.5 Systematic review guidelines 21

5 Discussion	23
6 Conclusion	27
7 Future Recommendations	28
References	29
Appendix A: Animal Experiments filters for PubMed and EMBASE	36
PubMed	*37*
EMBASE	*39*
Appendix B: Systematic Review filters for PubMed and EMBASE	42
PubMed	*43*
EMBASE	*43*
Appendix C: List of SR of AE included in this study	45
Appendix D: Assessment form	53
Appendix E: Step-by-step guide for effectively finding all relevant animal studies	55

Abbreviations

3Rs:	Russell and Burch's Three Rs (1959); Replacement, Reduction, Refinement
DOI:	Digital Object Identifier; unique character string used to identify a document
LAS:	Laboratory animal science
MeSH:	Medical Subject Heading; thesaurus terms used in MEDLINE database
PICOT:	Population, Intervention, Comparison, Outcome, Time
SR:	Systematic Review
SR of AE:	Systematic Review(s) of Animal Experiments

List of Figures

Figure 1: Flow chart of study selection process 17

Figure 2: Number of SR of AE published per year from January 2005 – July 2010. * 2010 was researched until 01 July 2010. 18

Figure 3: Frequency of the scores (0-18) for SR of AE between January 2005 and July 2010 (n=80) 18

Figure 4: Average quality score per year for systematic reviews of animal experiments (n=80) published between 2005 and 2010. Error bars represent s.e.m. * indicates a significant difference of $P<0.05$, ** indicates a significant difference of $P<0.01$. Significance was calculated using ANOVA followed by Least Significant Difference tests. 19

Figure 5: Overall score per assessment item of the SR of AE complete data set (n=80). DBN = reporting full database names, MDB = using multiple databases, OSE using other sources of evidence, Y= reporting years that were searched, SSR= repeatable search strategy, SSQ = quality of the search strategy, S = short explanatory summary, LR = reporting of language restrictions, D = reporting the search date 20

Figure 6: Overall score per assessment item of the SR of AE complete data set (n=80), split out in score categories (+, +/- and -). DBN = reporting full database names, MDB = using multiple databases, OSE using other sources of evidence, Y= reporting years that were searched, SSR= repeatable search strategy, SSQ = quality of the search strategy, S = short explanatory summary, LR = reporting of language restrictions, D = reporting the search date 21

Figure 7: Comparison of the average score of SR of AE (n=80) published in journals that refer to SR guidelines in their "Author guidelines" and those that do not. * Indicates a significant difference of $P<0.05$ 22

Figure 8: A: Number of studies published with and without guidelines with a score of > 14 pt. (n=10). B: Number of studies with and without guidelines with a score of <7 pt. (n=10). All studies were published between January 2005 and July 2010. 22

List of Tables

Table 1: Some chronologically listed developments in science and society that had an influence on the use of animals in research...5
Table 2: Essential reporting items for the search strategy of systematic reviews of animal experiments and the way they were assessed with +, +/- and - scores......................15
Table 3: Number of resulting citations for PubMed and EMBASE search strategy executed on 1 March 2010 for SR of AE between 1 January 2005 and 1 March 2010 16

List of Appendices

Appendix A: Animal Experiments filters for PubMed and EMBASE....................................... 36

Appendix B: Systematic Review filters for PubMed and EMBASE 42

Appendix C: List of SR of AE included in this study 45

Appendix D: Assessment form.. 53

Appendix E: Step-by-step guide for effectively finding all relevant animal studies 55

Acknowledgements

I would like to thank my supervisors, Dr Marlies Leenaars (3R Research Centre, Central Animal Laboratory, Radboud University Nijmegen Medical Centre) and Arnold de Vries DVM (HAS Den Bosch University of Applied Sciences), for their support and advice during my dissertation period. Marlies, your help was irreplaceble and allowed me to set clear goals and finish this dissertation to high standards. Furthermore I am very grateful you and Merel allowed me to publish this dissertation in two articles and linked my name to your presentation in Helsinki. Arnold, thank you for putting my research in perspective and showing me the bigger picture.

To my other colleagues (Prof Merel Ritskes-Hoitinga, Dr Rob de Vries, Dr Carlijn Hooijmans, Dr Jo Curfs, Judith van Luijk MSc, Joppe Tra BSc and Robbertjan Huijbregtse BSc) at the 3R Research Centre: thank you for the endless advice, proof reading, conversations and allowing me to develop myself in areas related to scientific research, such as writing grant applications and having me involved in some conference aspects. And ofcourse for all the fun!

Furthermore, I would like to thank Mrs Alice Tillema (Medical Library, Radboud University Nijmegen Medical Centre), Prof Gert Jan van der Wilt (Department of Epidemiology, Biostatistics and HTA, Radboud University Nijmegen Medical Centre), Dr Mariska Leeflang and Dr Lotty Hooft (both Dutch Cochrane Centre) and Dr Gerben ter Riet (Department of General Practice, Academic Medical Centre, University of Amsterdam) for commenting on my drafts and being co-author for my publications. Dr Kathleen Jenks (Department of Epidemiology, Biostatistics and HTA, Radboud University Nijmegen Medical Centre), thank you for the assistance with the statistical part of this dissertation.

Finally, I need to thank my parents. Without them I would not have started the MSc course. Mom and dad, you made this dissertation possible. I will be gone for three years to pursue a PhD, but I'll be back!

"Once thought, an idea cannot be unthought." (Edward de Bono, 2004)

Who ever knew this would be so much the case for the concept of systematic reviews of animal experiments?

1 Introduction

Laboratory animal science is a multidisciplinary field of study that aims to contribute to high quality research and to animal welfare (van Zutphen and Ohl, 2009). It is generally accepted that better laboratory animal welfare will lead to more valid test results (Baumans, 2005). The laboratory anmial science field tries to achieve better welfare and science by implementing Russell and Burch's "Three Rs", which were first proposed in their *The Principles of Humane Experimental Technique* (Balls, 2009). This work promotes the responsible use of animals in science.

Responsible use of animals comprises a number of things, such as looking for non-animal alternatives for an experiment, using less animals through better experimental design and statistical calculations (Howard *et al.*, 2009), but also choosing correct animal models in order to avoid the unnecessary, and thus unethical, use of animals (Gauthier and Griffin, 2005). Finally, if animals are to be used, procedures should aim to minimise discomfort to the animals (van Zutphen and Ohl, 2009). In this light, responsible use of animals is the same as implementing the Three Rs. Russell and Burch gave the following definitions for their Three Rs (Balls, 2009):

Replacement of animals by alternative, non-animal, methods
Reduction of the number of animals used in an experiment
Refinement of the experiment in order to minimise discomfort to the animal

Responsible use of animals in science, and thus Three Rs implementation, can be achieved in different ways. One of these can be performing a systematic literature review of existing animal studies (Knight, 2007).

In a systematic review (SR), all available relevant literature about a specific research question is identified, appraised, selected and ultimately extracted in order to generate new data (Bachmann *et al.*, 2003). These new results provide scientists with a better understanding and an evidence-based summary of the present situation concerning the particular research question (Guimaraes, 2009).

There are several steps to take when writing a SR (Pai *et al.*, 2004):
1. Compile a systematic research question

2. Design and execute a search strategy
3. Study selection: in- and exclusion of relevant studies according to predefined criteria
4. Quality assessment of the included studies
5. Data extraction
6. Data synthesis/meta-analysis
7. Discussion of the retrieved information
8. Formulating a conclusion based on the retrieved information

When a SR of AE is performed before a new animal experiment is conducted, researchers can determine whether the existing evidence justifies performing the experiment (Knight, 2008). Furthermore, when looking at a translational medicine context, performing a systematic review of animal experiments can help determine the right animal model and experimental design for an experiment (Peters et al., 2006) and improve predictability of a treatment (Pound et al., 2004; van der Worp et al., 2005; Perel et al., 2007), thus reducing the number of animals used and at the same time improve patient safety (Pound et al., 2004; Knight, 2007).

Even though SRs are being seen as the highest level of evidence in human (Horvath and Pewsner, 2004; Pai et al., 2004), and may reduce the number of animal experiments needed, they are not common practice in laboratory animal science. Research by Leenaars et al. (2009), Hooijmans et al. (2010b) and Cuijpers et al. (in prep.) showed that this might be caused by the fact that performing a SR of animal studies is not easy, because systematically searching the literature in order to retrieve all possibly relevant studies to a subject turned out to be troublesome, e.g. because scientists do not know how to efficiently use literature databases, even though they work with them every day (Hooijmans et al., 2010b).

Guidelines for writing SR of AE do not yet exist and should be developed. However, before this can be done an assessment of the current SR of AE should be made. Therefore, this study aims at assessing the literature search step in the SR writing process. The goal of this dissertation is to provide an assessment of the current quality of search strategies used in SR of AE, together with a comparison of the quality of these search strategies for SR of AE with the quality search strategies for human health care SRs.

This introduction forms chapter one. Chapter two is a literature review, looking at laboratory animal welfare, at the foundations of SR theory, at strengths and weaknesses of both human and animal SRs, as well as at potential opportunities SRs offer to the field of laboratory animal science. The third chapter explains the methods and materials used for this study. The results are displayed in chapter four and discussed in chapter five. Chapter six forms a conclusion, followed by future recommendations in chapter seven. All relevant background material is included in the appendices.

The purpose of this study is to aid in the introduction of systematic reviews in the field of laboratory animal science by investigating the quality of the search strategies used in current systematic reviews of animal experiments. Ultimately, this research should lead to an increase of laboratory animal welfare by increasing the quality of systematic reviews of animal experiments.

2 Literature Review

This review will discuss and illustrate the different aspects concerning the influence of search strategies of systematic reviews on animal welfare. Starting with an introduction in laboratory animal science and welfare, through the theory of systematic reviews and search strategies to a conclusive paragraph describing the influence a good search strategy has on a systematic review and why this is important for laboratory animal welfare.

2.1 Laboratory animal welfare

In the 18th century the idea of the capability of empirical science to improve human living conditions started to become generally accepted (van Zutphen and Ohl, 2009). It also became clear that, for medical and pharmaceutical science to progress, animal experiments were needed. However, in the 19th century opposition to animal experimentation also began to rise. In 1875 the Victoria Street Society was founded in the UK (Baumans, 2005). It was the first anti-vivisection movement in Europe and the UK. Under pressure from this movement, the UK Government accepted the Cruelty to Animal Act in 1876, which was the first law to protect animal used in research.

The use of animals in research increased due to breakthroughs in science (see Table 1). In 1859 Charles Darwin published *On the Origin of Species*, in which he defended the biological similarities between humans and animals (Darwin, 1859). In 1865 Claude Bernard published *Introduction a l'étude de la médecine experimentale*, in which he pleaded for the use of animals for the development of experimental medicine (van Zutphen and Ohl, 2009). Laboratory animals became an important substitute for human test subjects after Koch's Postulates were published in 1884, in which Robert Koch stated that proof for the pathogenicity of a microorganism is only valid after a successful inoculation and infection of healthy laboratory animals (van Zutphen and Ohl, 2009). After World War I the industrial production of pharmaceuticals and medical-biological research began to develop. This resulted in an exponential increase in the use of animals in research (Baumans, 2005).

Led by the increasing demand for high-standard animal models and the increasingly critical view towards the use of animals in research, laboratory animal science developed in the 1950s. This multidisciplinary branch of science aims at improving the quality of research in which animals are used, and at improving the welfare of these animals. In 1959 Russell and Burch published *The Principles of Humane Experimental Technique*, in which they pleaded for

Table 1: Some chronologically listed developments in science and society that had an influence on the use of animals in research

Year	Development
1859	Charles Darwin published *On the Origin of Species*
1865	Claude Bernard published *Introduction a l'étude de la médecine experimentale*
1876	Foundation of the Victoria Street Society
1884	Koch's Postulates were published
Post WW I era	Rise of industrial production of pharmaceuticals
1950s	Development of laboratory animal science
1959	Russel and Burch publish *The Principles of Humane Experimental Technique*

the use of the guiding principles of Replacement (of animals by non-animal methods), Reduction (of the number of animals used) and Refinement (of the procedures to decrease discomfort to the animal), also known as the Three Rs (3Rs) (Balls, 2009). This work promotes the responsible use of animals in research.

Where Refinement can be seen as improving welfare at the level of the individual animal, Replacement and Reduction improve welfare at population level, through ensuring that less animals suffer, because less animals are used in research. This interpretation of the 3Rs closely matches the idea of responsible animal use.

2.2 Systematic reviews

Over the last decades, the available amount of medical literature is increasing exponentially (Offringa and de Craen, 1999), in such a way that scientists are unable to keep up with the ever increasing amount of evidence they need in order to make decisions (Jadad and McQuay, 1996; Gerber *et al.*, 2007). Systematic reviews can help scientists by summarising large bodies of evidence (Jadad *et al.*, 2000; Horvath and Pewsner, 2004) and helping to explain differences among studies on the same question (Gerber *et al.*, 2007).

A systematic literature review involves the application of scientific strategies, in ways that limit bias, to the assembly, critical appraisal, and synthesis of all relevant studies that address a specific clinical question (Jadad *et al.*, 1998a; Moher *et al.*, 2009). A meta-analysis is a type of analysis that uses statistical methods to combine and summarize the results of several primary studies (Phillips, 2005). Meta-analyses can, but not necessarily have to, be a part of a SR (Moher *et al.*, 2009).

The concept of bias points to a false positive or false negative outcome of the systematic review (Sena *et al.*, 2010b). This false outcome can be caused by different sources of bias (Jüni *et al.*, 2001), which will be discussed in more detail in chapter 2.3. Systematic reviews can only minimise the risk of bias, since the occurrence of bias is practically unavoidable because of unpublished studies (Montori *et al.*, 2005a; Tricco *et al.*, 2009). However, when done properly, a meta-analysis will help indicate the occurrence of bias (Phillips, 2005; Sena *et al.*, 2010b). As long as a researcher acknowledges this bias and takes its existence into account when formulating conclusions, minimising the risk of bias by systematically and rigorously searching and selecting literature is sufficient. Minimising the risk of bias also requires searching the so-called grey literature: government publications, conference papers, books, etc. (Savoie *et al.*, 2003).

While systematic reviews are regarded as the strongest form of medical evidence, several authors have claimed that SRs have shown, and still show, deficiencies (Chalmers and Haynes, 1994; Jadad *et al.*, 2000; Jüni *et al.*, 2001; Moher *et al.*, 2007; Simera *et al.*, 2010), and that their reporting could be improved by a universally agreed upon set of standards and guidelines (Moja *et al.*, 2005; Moher *et al.*, 2007; Sampson *et al.*, 2008). Kilkenny *et al.* (2009) have shown that this is also the case with animal studies.

A study by Shojania *et al.* (2007) found that of 100 guidelines reviewed, 4% required updating within a year, and 11% after two years; this figure was higher in rapidly changing fields of medicine. Seven percent of systematic reviews needed updating at the time of publication.

According to Pai *et al.* (2004), there are several steps to take when writing a SR:
1. Compiling a systematic research question
2. Designing and executing a search strategy
3. Study selection: in- and exclusion of relevant studies according to predefined criteria
4. Quality assessment of the included studies
5. Data extraction
6. Data synthesis/meta-analysis
7. Discussing the retrieved information
8. Formulating a conclusion based on the retrieved information

Even though the theory of SRs has been developed with health care in mind, it has already been applied to other fields of science. These fields include e.g. veterinary medicine, social sciences and psychology. The last few years, SRs have also been written in the field of laboratory animal science (Mignini and Khan, 2006; Peters et al., 2006).

2.3 Designing a bibliographic search strategy

Finding all relevant literature is an essential step in writing a SR and should be a systematic process (Montori et al., 2005b). A literature search can be broadly divided into four phases. The first phase is always defining a systematically structured research question. The PICOT (Population, Intervention, Comparison, Outcome, Time) mnemonic provides a helpful tool (Hooijmans et al., 2010b).

For example, the research question "*Is there evidence of reduced allergenicity and clinical benefit of food hydrolysates in animals with cutaneous adverse food reactions, compared to non-hydrolysate food?*" (adapted from Olivry and Bizikova, 2010) forms a systematically structured research question: Population (animals with cutaneous adverse food reactions), Intervention (feeding food hydrolysates), Comparison (feeding non-hydrolysate food) Outcome (reduced allergenicity and clinical benefit), Time (not used) are present.

Once a research question has been developed, it can be divided into three sections that form the basis for the search strings: the Animal Studies filter (Hooijmans et al., 2010b) for finding all animal studies (Popuation) and Intervention and Disease filter, in this case "food hydrolysates" and "adverse food reactions".

According to Hart et al. (2005), the next phase is to decide what sources to use for a search. It is advisable to search in at least two digital databases (Woods and Trewheellar, 1998; Assendelft et al., 1999; Pai et al., 2004). In addition, selected references in the articles found in the databases should be evaluated as well, since they can provide new evidence, but also more insight into the field of study (Assendelft et al., 1999). So-called 'grey literature' (books, conference papers, government publications) should also be searched, in order to find unpublished studies and minimise the risk of bias (Yoshii et al., 2009; Sena et al., 2010a; van der Worp et al., 2010).

In the third phase, search strings are developed that are specifically designed for the databases intended to be used. When searching bibliographic databases for all possibly relevant animal studies, the structure of the search string is essential (Sampson and

McGowan, 2006; Krupski et al., 2008; Yoshii et al., 2009). Simple *ad random* entering of search terms in the search field is not enough. It is important to search for synonyms of search terms, take note of different spellings (e.g. US English and UK English) and use both singular and plural terms. Key factors to consider are the use of thesaurus terms (Chilov et al., 2007) (see e.g. the MeSH database for Pubmed and Emtree terms for EMBASE), and the use of search field specifications (title, abstract, MeSH/Emtree, journal etc.) (Haynes et al., 2005).

The search string designed by Hooijmans et al. (2010b) can be copied into PubMed and used as a search set to find all animal studies. The disease set and intervention set have to be built by carefully choosing MeSH terms and keywords for title and abstract and connect these with the Boolean operator AND (Sampson and McGowan, 2006).

The use of the search string designed by Hooijmans et al. (2010b) returns a large number of citations, some possibly more relevant than others. It is however very important to include as much relevant citations as possible in a SR to get as much information possible about the question, to be complete and to minimise the risk of bias (Sena et al., 2010b). In order to help minimise this risk of bias it is imperative to use at least two bibliographic databases. Both in humane studies (Woods and Trewheellar, 1998; Wong et al., 2006) and animal experiment related studies (Hooijmans et al., in preparation; de Vries et al., in preparation) additional unique studies were found when using EMBASE next to PubMed to retrieve biomedical literature, showing the need for using more than one bibliographic database.

Language restrictions should not be used. Studies with a neutral or negative result are more often published in a non-English language (Moher et al., 1996; Egger et al., 1997; Chalmers and Matthews, 2006). This will lead to a false positive outcome of the study due to language bias (Grégoire et al., 1995; Egger et al., 1997; Moher et al., 2003). Even though a more recent study claims the effect of language restriction is little (Jüni et al., 2002), it also acknowledges the upcoming importance of studies in Chinese and recognises that the effect of language bias is largely dependent on the field of study.

The last database-specific step before executing the search would then be to combine the Animal Studies set with the Disease set and the Intervention set, which is done by connecting the corresponding set numbers with the Boolean operator AND. Experience shows that the first search action will most likely return a large number of irrelevant results (de Vries et al., in prep.; Hooijmans et al., in prep.). Results can be improved by adapting the search string and evaluate the keywords and thesaurus terms used for their efficacy (Sampson and McGowan, 2006).

When an optimal balance between relevant and irrelevant articles is reached, the search string can be translated to another database environment. Because the indexing practices vary and thesaurus terms do not necessarily have equivalents in both databases, it is unfortunately not possible to directly translate search strings from one database to another (Bachmann et al., 2003; Wong et al., 2006). Translating therefore asks for careful consideration and re-evaluation of the search string. However, the principles for designing search strings remain the same, since these principles are not dependent on the database language.

The last two phases to obtain all potentially relevant references without duplicates are executing both the final PubMed and EMBASE searches and cross referencing the results in order to delete double citations. This can easily be done through the use of bibliographic database software, like e.g. EndNote or RefWorks.

2.4 Reporting a search strategy

Equally important to reproducibility is the methodology for reporting of the search strategy. Transparent reporting of the complete search strategy that was used to retrieve existing knowledge facilitates repeatability (Sampson and McGowan, 2006; Hooijmans et al., 2010a; Kilkenny et al., 2010) and allows the quality of the SR to be determined by others (Jadad et al., 1998b). This increases the reliability, utility and impact of the SR (Hooijmans et al., 2010b; Kilkenny et al., 2010; Simera et al., 2010) and minimises the risk of bias (Khan and Mignini, 2005; Chalmers and Matthews, 2006; Higgins and Altman, 2009; Macleod et al., 2009; van der Worp et al., 2010). However, inadequate reporting is a common problem in systematic reviews of human health care topics (Sampson and McGowan, 2006; Shea et al., 2006).

There are a number of important items that need to be reported when describing a search strategy (Sampson et al., 2008; Yoshii et al., 2009). Reporting the full names of databases in which was searched, together with the full names of other sources of evidence used, enhances repeatability and transparency. The same goes for precisely reporting the years to which the search was limited, or in case no limit was used, the years that the used databases comprise. Reporting the full electronic search string for each database used will also increase repeatability and transparency, while a short descriptive summary of the search string will enable a quick understanding of the context in which the string was developed. As mentioned before, it is strongly advised not to include language restrictions in the search string. In case no restrictions on language have been used, this fact should still be reported (Yoshii et al., 2009).

Since optimal use of bibliographic databases requires an understanding of the underlying principles and techniques of database software, content and Boolean operators, the help of a (library) information specialist is indispensible (Sampson and McGowan, 2006; Yoshii et al., 2009). Authors of systematic reviews of animal experiments should consider seeking advice from such specialists after developing search strings. Library information specialists are trained in database searches and are able to advise on improvements.

2.5 The influence of a literature search strategy on systematic reviews and laboratory animal welfare

Peters et al. (2006) state that SRs offer a structured and transparent approach to searching, reviewing and evaluating all available relevant evidence and so are directly relevant to the movement towards the 3Rs in animal research. Furthermore, the increased precision of a meta-analysis over a single study in particular has implications for reduction of animal research. Some animal experiments provide little reliable information as a result of striving to use as little animals as possible (Guimaraes, 2009). According to Roberts et al. (2002), Pound et al. (2004) and Sena et al. (2007), this is were a SR including a meta-analysis would increase the precision of the estimates of treatment effect, e.g. rather than to conduct another experiment because the previous experiments did not have sufficient power, it might be appropriate to conduct a meta-analysis of the existing evidence. Meta-analyses also offer the ability to explore consistency and generalisability of effects, and a framework for investigating heterogeneity between studies and possible publication bias (Phillips, 2005; Peters et al., 2006). Through an understanding of the different sources of bias that may be apparent in primary studies, the quality of conducting and reporting animal experiments may be improved, since the understanding of bias facilitates prevention of bias (Sena et al., 2010b).

New insight can be gained, without using animals, by performing detailed analyses of existing evidence (Knight, 2007, 2008). This new insight may inform a researcher that his experiment is not justified and needs rethinking and redesign (Festing, 2004; Lemon and Dunnett, 2005). According to (Macleod et al., 2005; Wood and Hart, 2007), systematic reviews of animal experiments can also contribute to laboratory animal welfare by allowing researchers to determine the best animal model for their experiment. Existing evidence may rule out certain animal models, or favour others (Bracken, 2009). This leads to less animals being used, since experiments will not be performed in models that are not suitable for the experiment of have no value for the outcome (Lindl *et al.*, 2005). Futhermore, SR of AE prevent allow researchers to perform their experiments in the best model straight away, instead of first using a cheap and easy model, before repeating the experiment in more expensive, larger, animal models (Wood and Hart, 2007).

Furthermore, SRs also have a role in translational medicine. Animal experiments usually form the basis of risk assessments for safe human exposure limits to new drugs, chemical substances in the environment, in food and in commercial products (Baumans, 2005). In such cases, human evidence is often limited and results from animal experiments are the main source of evidence (Hackam and Redelmeier, 2006; Hackam, 2007). It is known that results from animal studies do not always, if not seldomly, have the desired effect in man (Roberts *et al.*, 2002; Pound *et al.*, 2004; Knight, 2007; Macleod *et al.*, 2008; Sena *et al.*, 2010a; van der Worp *et al.*, 2010). Moreover, cases exist where animal experimental results were used and products tested in humans, resulting in negative effects. A SR performed afterwards showed that based on the experimental results, the product should not have been tested in humans (Pound *et al.*, 2004). In order to help resolve this problem, SRs should be performed of all evidence, including animal research, before proceeding to clinical trials (Wheble *et al.*, 2008).

One of the weaknesses of SRs is that they are dependent on high quality primary studies, with clear and complete reporting (Kilkenny *et al.*, 2009; Hooijmans *et al.*, 2010a). However, until recently, no clear guidelines for reporting animal studies were available. This results in SRs that are of less quality. Recently, Hooijmans *et al.* (2010a) and Kilkenny *et al.* (2010) published guidelines for reporting animal experiments. This should lead to better primary studies and increase feasability of SR of AE (Hooijmans *et al.*, 2010a). In order to facilitate SR of AE even more, guidelines for writing these would be helpful. These guidelines are currently being developed, starting with a step-by-step guide for searching all potentially relevant animal studies (van Veggel *et al*, in prep., see appendix E)

In order to write a high quality SR of AE, one needs to find all available relevant evidence. The more evidence is compiled in a SR of AE, the better the SR will be and the larger the influence of the SR on animal welfare is. Therefore, an efficient and effective search strategy will help improve laboratory animal welfare.

3 Materials and Methods

3.1 Search strategy for identification of studies

The MEDLINE (PubMed platform) and EMBASE (OvidSP platform) databases were queried using the 'Animal experiments' filter by Hooijmans *et al.* (2010b) (see appendix A) combined with a 'Systematic review' filter (see appendix B) on 1 March 2010. Only citations referring to the publication type "review" were used. This study follows up on a study by Peters *et al.* (2006) and Mignini *et al.* (2006), who have looked at SR of AE until January 2005. Therefore, the publication years were limited from 1 January 2005 to 1 March 2010. All citations were imported into an EndNote X3 (Thomson Reuters) library, after which duplicates were removed automatically and manually. A second search, identical in methodology, was executed on 1 July 2010, in order to retrieve SRs of AE from 1 March 2010 to 1 July 2010. Results from this second search were added to the existing first database.

3.2 Study selection

The initial study selection (see Figure 1, p. 18, for a flow chart) was done based on title and abstract. The initial set of citations was screened by hand and all studies that did not involve animal models, studies or experiments were excluded. Studies that both contained human and animal data were included. The citations were then filtered using the search function offered in the Endnote X3 program. All citations with the title or abstract sections containing the keywords "systematic review", "Cochrane", "systematically review" "meta-analysis", "meta analysis", "systematic literature review", "data source", "systematic search", "systematic literature search" or "evidence synthesis" were included. A second filter was applied, filtering all studies in the Chinese language. These studies were saved in a separate database for future analysis, but were excluded for this study. The remainder of the citations were screened manually for studies that were missed by the automatic searches.

A second, manual, study selection phase was performed based on full text articles. Studies not involving animal models and those that were not systematic reviews were excluded. In order to qualify as a SR, studies had to be structured systematically. Studies that contained both human and animal data were excluded in cases were the animal and human data were inseparable. Full-text articles and URLs to full text articles were automatically retrieved through the automated full text search function in the EndNote X3 program. The other articles were retrieved manually through the Scopus website, publisher's websites and searches of the Radboud University Nijmegen Medical Centre Medical Library, as well as by contacting authors directly.

The study selection phase was performed by one author. In case of uncertainty on whether to in- or exclude a study, a second reviewer was consulted.

3.3 Quality assessment

An assessment system was developed based on the Cochrane Handbook for Systematic Reviews of Interventions (Cochrane Collaboration, 2009), on the PRISMA Statement (Moher et al., 2009) and on studies by Mignini et al. (2006), Peters et al. (2006) and Yoshii et al. (2009). From these publications, items considered essential for (human) SRs were retrieved (see Table 2).

This list of items was then converted into an assessment system (see Table 2) and a reporting form was developed (see Appendix D). When an item was completely present and reported, it was awarded "+", the partial presence and reporting of an item was awarded "+/-" and not reporting or not executing an item was awarded "-". This system allowed a study to accumulate a maximum of 9 "+". For statistical purposes, every "+" was converted to the numerical value 2, every "+/-" to a value of 1 and every "-" to a value of 0. This way, the maximum score for a study became 18.

3.4 Data extraction

All included studies were masked by deleting all references to authors, journal title, publication year, DOI-number and author affiliations. In an Excel (Microsoft®) file the studies were randomly matched to a number. All articles were printed and masked and the corresponding number was applied, so that afterwards the numbers could be matched again to the corresponding article for analysis purposes.

The data extraction was performed independently by two reviewers. Differences were resolved through discussion. When discussion did not resolve the issue, a third reviewer was consulted.

3.5 Statistical analysis

All statistical calculations were performed with SPSS Statistics version 16 (IBM® SPSS®) in a Windows XP (Microsoft®) environment. The dataset was explored using the Explore function of SPSS and was found normally distributed. ANOVA was used to calculate significant differences. Least Significant Difference (LSD) tests were used for pair-wise comparisons. Since the dataset was small, no adjustment for multiple comparisons was applied.

3.6 Developing a step-by-step search and reporting guide

Guidelines originating from human health sciences formed the basis for a search guide for animal studies, such as the Cochrane Handbook for Systematic Reviews of Interventions (Cochrane Collaboration, 2009), studies by Sampson and McGowan (2006), Sampson et al. (2008) and Yoshii et al. (2009), and the QUOROM (Moher *et al.*, 1999), PRISMA (Moher *et al.*, 2009), CONSORT (O'Connor *et al.*, 2010; Schulz *et al.*, 2010) and REFLECT (O'Connor *et al.*, 2010) reporting guidelines. A library information specialist from the Medical Library of the Radboud University Nijmegen Medical Centre was consulted for expert advice.

The guide has been developed together with scientists experienced in writing systematic reviews and has been tested by a library information specialist. A draft of this guide is included as Appendix E and has aready been used in academic education.

Table 2: Essential reporting items for the search strategy of systematic reviews of animal experiments and the way they were assessed with +, +/- and - scores

Abbreviation	Essential item	Score
MDB	Searching multiple databases	+ more than one database used +/- one database used - no databases used or reported
DBN	Reporting the full name of all databases used	+ full names of all databases reported +/- partial names reported or not all databases named - no database names reported
OSE	Using other sources of evidence	+ more than one alternative source used +/- one alternative source database used - no alternative source used or reported
D	The date the search was performed	+ full date reported (day, month and year) +/- partial date reported - date not reported
Y	The years limiting the search	+ lower and upper year limit reported +/- lower of upper year limit reported - years not reported
SSR	Repeatability of the search strategy	+ fully repeatable +/- partially repeatable - not repeatable
SSQ	Technical quality of the search strategy	+ high technical qulaity +/- intermediate technical quality - low technical quality
S	Short summary of the search strategy	+ summary present - summary absent
LR	Reporting the use of language restrictions	+ use of language restriction reported - use of language restrictions not reported

4 Results

4.1 Literature search

Applying the 'Animal experiments' filter (Appendix A) retrieved 4,739,212 studies for PubMed and 3,001,846 for EMBASE (Table 1). When this filter was combined with the 'Systematic Review' filter (Appendix B), 7,742 and 6,402 references were left respectively. Next, searching within these results for 'review' type pulications from 1 January 2005 - 1 March 2010 resulted in 2,428 citations for PubMed and 1,784 citations for EMBASE. After removing duplicates (articles occurring more than once in each database), and combining the results for each database 4,207 citations were left. A second removal of duplicates (citations occurring in both databases), this literature search strategy resulted in 3,608 potentially relevant studies.

Table 3: Number of resulting citations for PubMed and EMBASE search strategy executed on 1 March 2010 for SR of AE between 1 January 2005 and 1 March 2010

Step		PubMed	EMBASE
1.	'Animal experiments' filter[1]	4,739,212	3,001,846
2.	'Systematic review' filter[2]	131,750	96,858
3.	(1 AND 2)	7,742	6,402
4.	3 AND (Publication type = 'review')	4,619	2,983
5.	4 AND (Publication date = 2005/01/01 - 2010/03/01)	2,428	1,784
6.	Remove duplicates	2,427	1,780
7.	Combination PubMed and EMBASE	4,207	
8.	Remove duplicates	3,608	

[1]See Appendix A
[2]See Appendix B

The second literature search, executed on 1 July 2010 and searching for SR of AE between 1 March 2010 and 1 July 2010, resulted in another 111 citations for PubMed and 215 for EMBASE.

4.2 Study selection

After in- and exclusion (see Figure 1), 80 full-text systematic reviews of animal experiments were left, which were used for this study (see appendix C for a complete list of references).

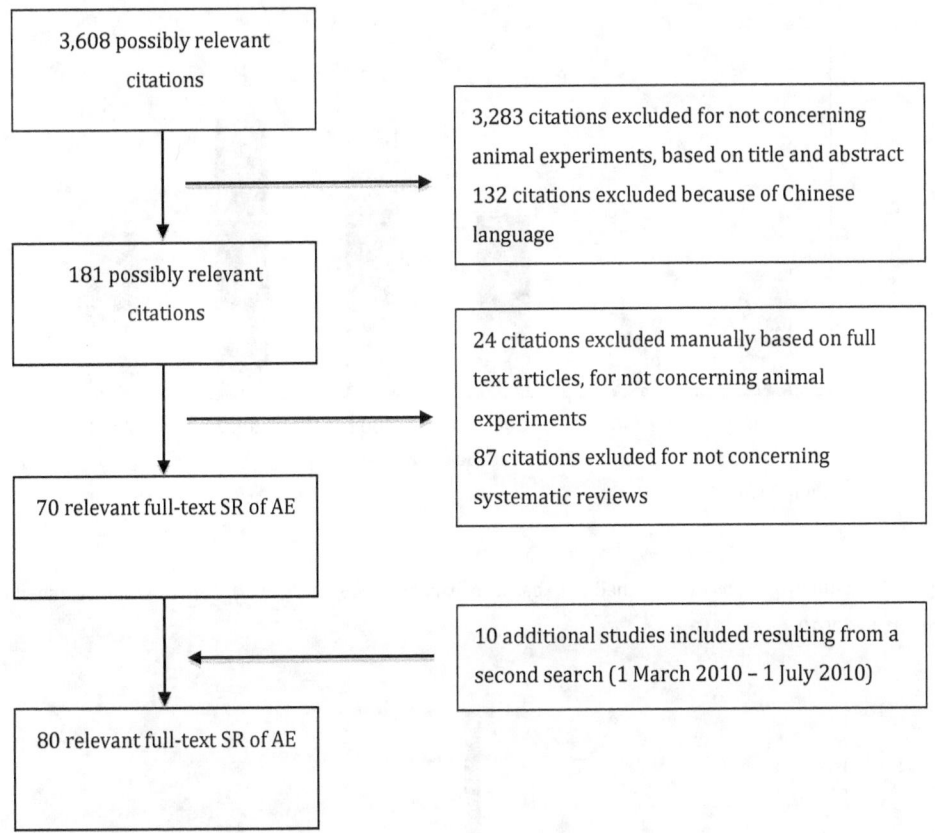

Figure 1: Flow chart of study selection process

4.3 Publications per year

The number of SR of AE published per year appears to increase (Figure 2), with nine SR of AE published in 2005, six in 2006, 13 in 2007, 20 in 2008, 20 in 2009 and 12 between January and July 2010. It is expected that the increasing line continues, resulting in n>20 for the whole of 2010.

After evaluating the scores that were given, the score with the highest frequency was 9 points, and the average score was 10,26 points. The minimum score (0 points) and the maximum score (18 points) were not given (see Figure 3). The lowest score given was 5 points, the highest score given was 16 points.

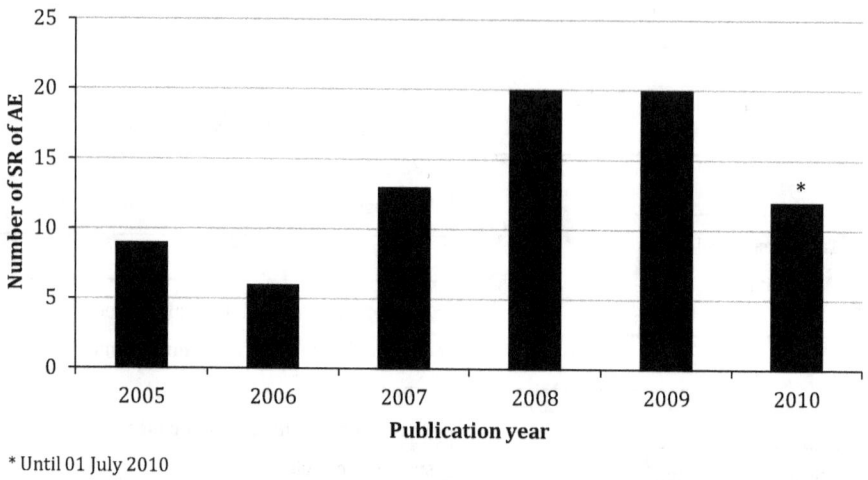

*Until 01 July 2010

Figure 2: Number of SR of AE published per year from January 2005 – July 2010. * 2010 was researched until 01 July 2010.

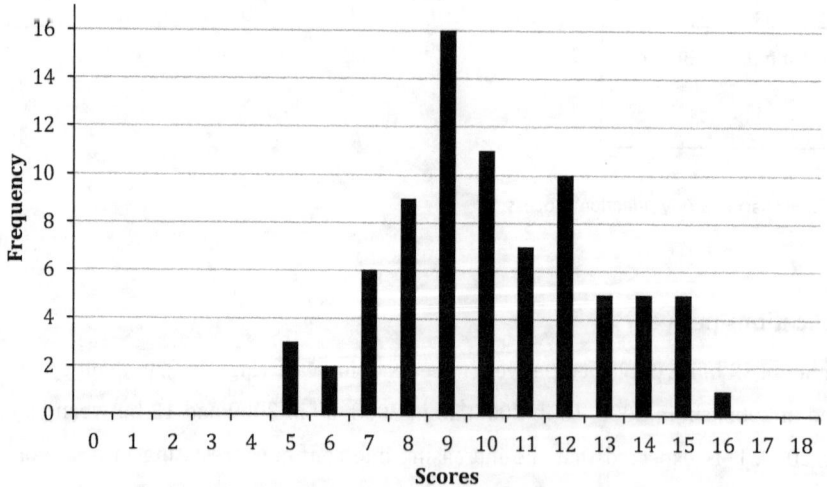

Figure 3: Frequency of the scores (0-18) for SR of AE between January 2005 and July 2010 (n=80)

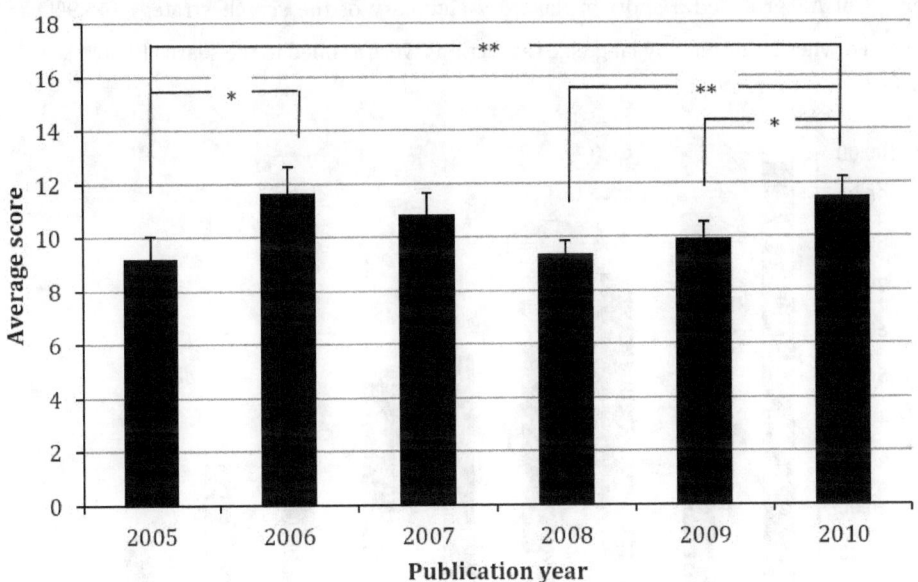

Figure 4: Average quality score per year for systematic reviews of animal experiments (n=80) published between 2005 and 2010. Error bars represent s.e.m. * indicates a significant difference of $P<0.05$, ** indicates a significant difference of $P<0.01$. Significance was calculated using ANOVA followed by Least Significant Difference tests.

When looking at the average quality scores per year (Figure 4), the results, an average score of 9.2 points for 2005, 11.7 points for 2006, 10.8 points for 2007, 9.4 points for 2008, 10.0 points for 2009 and 11.5 points for 2010, show an overall significant difference (ANOVA, $P<0.05$) After applying a Least Significant Difference test, a significant increase of quality between 2005 and 2010 ($P<0.01$) was found. Furthermore, a significant increase was seen between 2008 and 2010 ($P<0.01$) and between 2009 and 2010 ($P<0.05$).

4.4 Assessment items

When looking at the individual items that were scored (Figure 5) for the complete dataset of 80 SR of AE, it is apparent that not all items score evenly well. Almost all SR of AE reported the full names of all the databases used (98.1%), used more than one literature database (88.9 %) and supplemented their evidence with other sources of evidence (73.1%). Little over half of the studies reported the years the search encompassed (57.5%) and reported their search strategy in a high quality (51.9%) and repeatable way (55.0%). Less than half of the SR of AE provided a short explanatory summary of the search strategy (46.9%) and reported whether or not any language restrictions were applied in the search (30.6%).

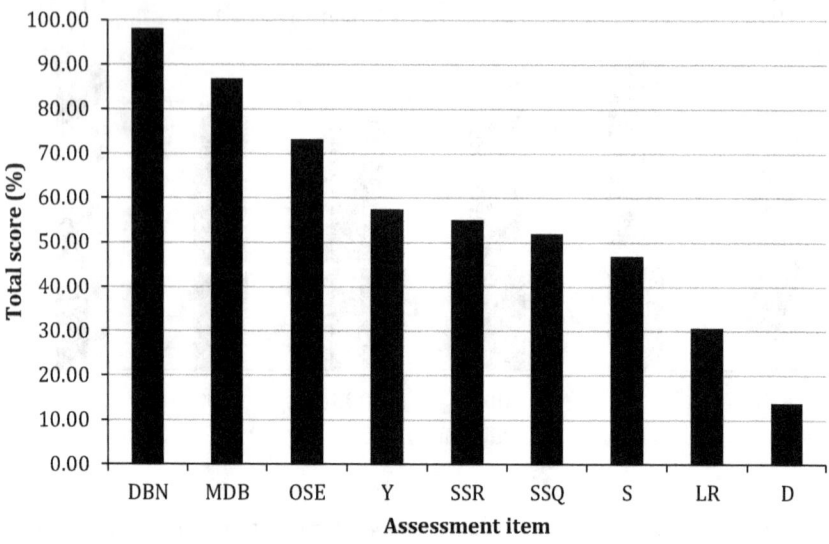

Figure 5: Overall score per assessment item of the SR of AE complete data set (n=80). DBN = reporting full database names, MDB = using multiple databases, OSE using other sources of evidence, Y= reporting years that were searched, SSR= repeatable search strategy, SSQ = quality of the search strategy, S = short explanatory summary, LR = reporting of language restrictions, D = reporting the search date

Almost none of the SR of AE reported the date the search strategy was executed (13.8%).

In contrast, when looking at the individual assessment items in more detail (Figure 6), it becomes clear that the quality of the search strategies can be improved tremendously. Nearly all SR of AE scored a + on fully reporting all names of the databases that were used (96.3%). Most of the SR of AE used more than one bibliographic database (75,0%), some studies reported using only one database (23.8%) and a small number of SR of AE did not report the use of databases (1.2%). More than one alternative sources of evidence were used by 60% of the SR of AE, while 26.3% reported using one alternative source of evidence and 13.7% did not use other sources of evidence. Less than half of the SR of AE reported both the years between which literature was searched (43.8%), 27.5% only reported the cut-off year and 28.8% of SR of AE did not report the years between which was searched. Only 30% of the assessed SR of AE contained a transparent and repeatable search strategy, while 50.0% had a search strategy that was partly repeatable and 20.0% contained a search strategy that was not repeatable. The technical quality of the search strategy did not differ much, with 20% of

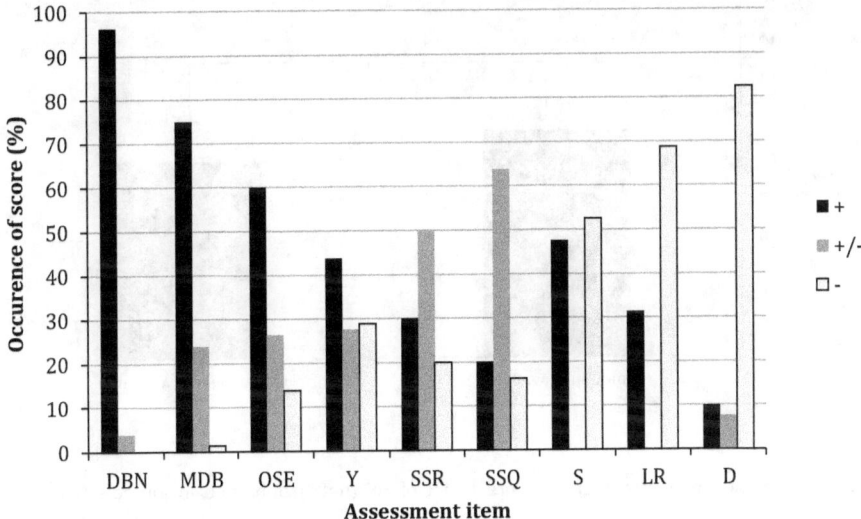

Figure 6: Overall score per assessment item of the SR of AE complete data set (n=80), split out in score categories (+, +/- and -). DBN = reporting full database names, MDB = using multiple databases, OSE using other sources of evidence, Y= reporting years that were searched, SSR= repeatable search strategy, SSQ = quality of the search strategy, S = short explanatory summary, LR = reporting of language restrictions, D = reporting the search date

the publications being of good quality, 63.8% being of intermediate quality and 16.2% being of low quality. The short explanatory summary was present in nearly half of the SR of AE (47.5%) and absent in little over half of the publications (52.5%). 31.3% of the SR of AE reported whether or not language restrictions were used, 68.7% did not report the use of language restrictions. The date the search strategy was performed was completely reported in 10.0% of the SR of AE, partially reported in 7.5% of the publications and not reported in 82.5%.

4.5 Systematic review guidelines

Figure 7 shows the average quality score of SR of AE that are published in journals that either provide or do not provide a reference to guidelines for writing SRs. SR of AE published in a journal that provides SR guidelines show a significantly higher quality score than those published in journals which do not provide guidelines (*P*<0.05). The difference, 1.5 point, is however only small.

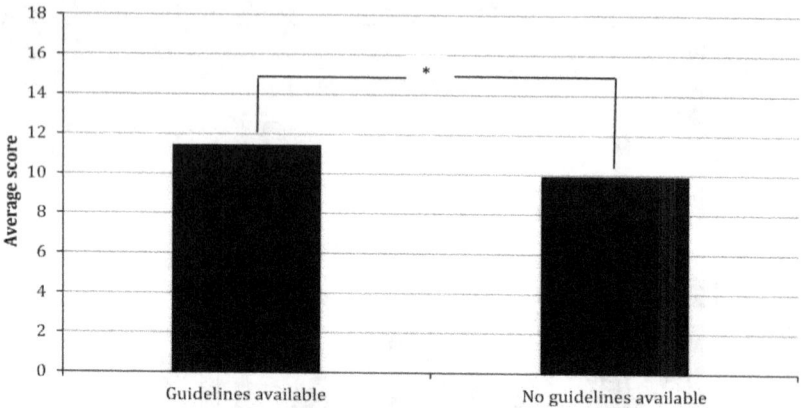

Figure 7: Comparison of the average score of SR of AE (n=80) published in journals that refer to SR guidelines in their "Author guidelines" and those that do not. * Indicates a significant difference of *P*<0.05.

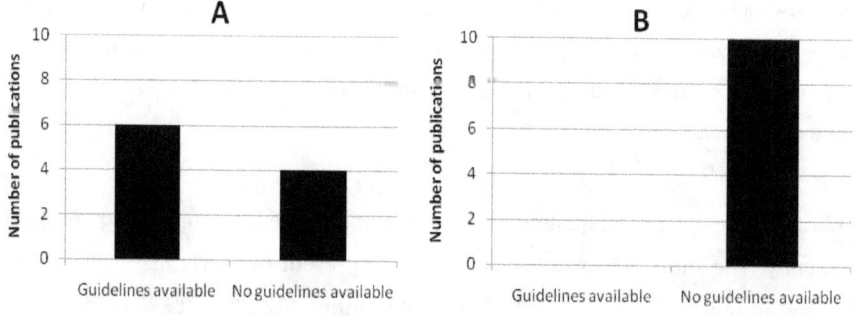

Figure 8: A: Number of studies published with and without guidelines with a score of > 14 pt. (n=10). B: Number of studies with and without guidelines with a score of <7 pt. (n=10). All studies were published between January 2005 and July 2010.

When looking at SR of AE scoring more than 14 points (A, n=10) and less than 7 points (B, n=10), none of the SR of AE scoring less than 7 points were published in journals that provide reference to guidelines for writing SR of AE in the "Guidelines to authors" section of their website. In contrast, 60% of the SR of AE scoring more than 14 points was published in journals providing those guidelines (see Figure 8).

5 Discussion

The inclusion of the search string designed by Hooijmans *et al.* (2010b) in the applied search strategy returns a large amount of citations, some possibly more relevant than others. It is however very important to include as much relevant citations as possible to inform the review question (Bachmann *et al.*, 2003) and in order to be complete and to minimise the occurrence of bias (Sena *et al.*, 2010b). Therefore the author advises to partly sacrifice specificity for sensitivity, albeit that authors should always keep in mind whether or not the extra effort involved with broadening the search and discarding more citations will lead to more relevant citations.

The same principle applies to the use of multiple databases: even though using more databases might lead to more *possibly* relevant citations, this does not automatically lead to more *actually* relevant citations. However, in order to minimise the risk of bias it is imperative to use at least two bibliographic databases (Macleod *et al.*, 2005; Yoshii *et al.*, 2009; Sena *et al.*, 2010b). When looking at PubMed and Embase, the two most commonly used biomedical literature databases, there are quite a number of citations that occur in both databases (Bachmann *et al.*, 2003). However, Embase does provide a relevant number of unique citations and is therefore a valuable source of evidence (Woods and Trewheellar, 1998; Wong *et al.*, 2006).

For this study, a large number of publications written in the Chinese language were excluded because of lack of resources. While in an ideal situation these publications should be included and translated, this was not possible in this project. This does however increase the risk of bias, since excluding publications based on publication langage causes language bias (Egger *et al.*, 1997; Jüni *et al.*, 2002). As a result, the outcomes of this study might change, either positively of negatively, when the excluded Chinese articles would the taken into account. Therefore, further research including all languages is warranted.

At the time of writing, no scoring systems existed for assessing the quality of systematic reviews of animal experiments. However, since both medical and animal experimental SRs rely on the same principles of thorough, systematic review of all available literature, it is only logical to assume that SRs for both fields can be assessed in an identical way. For this reason, the items considered important for search strategies for Cochrane reviews (Yoshii *et al.*, 2009), which are the highest quality reviews available, were used in this study and converted in to an assessment system. Shea *et al.* (2006) found that methodological quality assessment of SR is still in its infancy, and that substantial improvements can be made. This is also true for SR of AE, therefore the author does acknowlegde that the scoring system used in this study can and should be refined in the future. However, for the purpose of the current study, gaining a first insight in the quality of current SRs, the assessment system proved to be a clear and workable system.

One other point of discussion concerning the applied assessment system would be the lack of differentiation between studies that only partially fullfill all assessment items, or those that completely fullfill part of the items and not report other items. When looking at total scores, this gives a skewed result, since for example scoring 5+ and 5- leads to the same score as scoring 10 +/-. The author has tried to accommodate this skewness by providing detailed insight in the obtained scores (see Figure 6), but acknowledges this issue as a weakness in the applied assessment system. However, for the goal of this study, which is to provide a first insight in the quality of search strategies of existing SR of AE, the weakness is not believed to have a significant influence.

The goal of this study was not to determine which item of a search strategy was more important than the others, but to get an overall insight in quality. Therefore, the scoring system used treated all items identically. One can argue that certain items, like searching multiple databases, are more important than e.g. reporting the date the search was executed. However, in order to minimise the risk of bias and to give the reader as much information on methodology as possible, authors should include all decisions and steps taken (Higgins and Altman, 2009; Yoshii *et al.*, 2009). Each item has its own importance, and contributes to the overall goal of a SR.

Compared to human systematic reviews, the assessed SR of AE are of lower quality than their human medicine couterparts. Yoshii *et al.* (2009) looked at Cochrane reviews, and found that 100% of the studies that were assessed used more than one bibliographic database and reported all database names, as opposed to 75.0% and 96.3% for SR of AE. 91% reported the full date of the years covered by the search (43.8% for SR of AE), 88% reported the full search strategy for at least one database (30% for SR of AE), 69% reported the use of language restrictions, 26% reported a short descriptive summary of the search strategy (47.5% for SR of AE) and 11% reported the date the search was executed (10.0% for SR of AE). Even though solid evidence does not exist yet, the fact that authors have free online access to the *Cochrane Handbook for Systematic Reviews of Interventions* (Cochrane

Collaboration, 2009), which is a comprehensive guide on how to write Cochrane reviews, might explain the difference in quality. When guidelines for SR of AE become wideliy accepted and used, further research can point out whether the differences decrease.

While word count or page count have always been a limiting factor for authors when trying to publish a study, most publishers provide online space for extra appendices these days (Sampson *et al.*, 2008). This indicates that there is no reason anymore for not completely reporting search strings for all databases used. As long as authors report the availability of online appendices, and the documents are accessible, the reporting will still be complete and transparent and allow for the risk of bias to be minimised.

When looking at the number of SR of AE publications per year over the last five years, an increase can be seen. This increase in number of SR of AE being published is expected to continue. When we compare the development of SR of AE publication numbers with that of human SRs, both fields show a slow start, but when the concept of SR becomes generally accepted, the number of SRs will increase exponentially (Green *et al.*, 2009; Starr *et al.*, 2009). In order to see this effect in SR of AE, a further, future, review is needed.

In contrast to the number of publications, the quality scores for SR of AE over the years does not show a clear increase. However, the period 2008 – 2010 does show an improvement in quality that is expected to continue should clear guidelines become available. In human medicine, the quality of SRs, and that of the search strategies used, has been a stable factor, even though improvements can and should be made (Altman, 2004; Sampson and McGowan, 2006). The availability of clear guidelines has had its influence (Chalmers and Haynes, 1994; Gerber *et al.*, 2007), which is also part of the reason human SRs are of a higher standard than SR of AE (Mignini and Khan, 2006). One should consider though that development of human medicine SRs started over 20 years ago (Starr *et al.*, 2009), while performing SR of AE is a relatively recent and unknown method.

The peak in search strategy quality for the year 2006 might be explained by the fact that the CAMARADES group, a multi-institutional research group focussing on systematic reviews of treatments for ischaemic stroke, published quite a number of SR of AE in that year. Further research should attempt to find out whether CAMARADES-like knowledge groups have a positive influence on the quality of SR of AE.

Whether or not journal publishers would refer to SR guidelines in the "Authors guidelines" section of their website did make a difference. SR of AE published in journals that did refer to guidelines tended to have a significantly higher score than those published in a journal that did not refer to guidelines. The difference in quality score was however small. The peer-review process of journals might have an influence on this fact: when a peer-reviewed manuscript is sent back to the author with the advice to follow SR guidelines, the quality of the SR is expected to improve. As a result, authors are referred to guidelines, even though these are not mentioned on the website of the publisher. However, since a significant difference is still found, referring to SR guidelines does offer an opportunity to improve SR of AE quality. It would be good to investigate whether SR of AE that followed SR guidelines are accepted for publication sooner than SR of AE that did not follow SR guidelines.

Since optimal use of bibliographic databases requires an understanding of the underlying principles and techniques of database software, content and Boolean operators, the help of a (library) information specialist is indispensible (Sampson and McGowan, 2006; Yoshii *et al.*, 2009). Authors of systematic reviews of animal experiments should therefore seek advice from such specialists after developing search strings. Library information specialists are trained in database searches and are able to advise on improvements (Yoshii *et al.*, 2009). This will ensure that the systematic review of animal experiments will be of high quality.

Even though there is no direct link between high quality SR of AE and better animal welfare, evidence does suggest that SR of AE help to make sure that animal studies do not set out to answer questions that can be answered by looking at existing knowledge (Roberts *et al.*, 2002; Pound *et al.*, 2004; Guimaraes, 2009). This fact on itself already provides a strong ethical reason to promote and write SR of AE (Knight, 2008), let alone the fact that evidence suggests that performing systematic reviews, either human or animal, improves patient safety (Sena *et al.*, 2007).

6 Conclusion

High quality systematic reviews of animal experiment have a great potenial to benefit laboratory animal welfare through providing new evidence from existing data. This allows evidence-based decisions to be made regarding the justification of whether or not to perform an experiment and the design of that experiment, as wel as the correct animal model that should be used. These evidence-based decisions will reduce the number of experiments, and the number of animals used in experiments. This relates directly to Russell and Burch's Three Rs and to the responsible use of animals in science.

High quality systematic reviews start with a good research question and a thorough literature search strategy. Even though the quality of search strategies of systematic reviews of animal experiments shows improvement over the years, further improvements, both in quality and reporting, are still a great necessity. The risk of bias can only be minimised by complete and transparent reporting of all actions taken, and only when bias is minimised and taken into account will a study provide reliable results.

In the end, only the use of animals in experiments that generate reliable results can be ethically justified. Therefore, systematic reviews of animal experiments not only have scientific value, but also provide a way of meeting ethical requirements set by Animal Ethics Committees.

Referring to systematic review guidelines improves the quality of search strategies of SR of AE. This points out the need for journals to refer to guidelines, but also the need for specific SR of AE guidelines. The availability of clear guidelines will facilitate performing systematic reviews of animal experiments and contribute to the consistency of their quality.

7 Future Recommendations

For the near future, it will be crucial to have a central register for animal experiments, as is already in existence for human clinical trials (e.g. http://clinicaltrials.gov). This way, finding all potentially relevant animal studies becomes more feasible, since authors will have knowledge of unpublished results. This will, in turn, decrease the risk of unnecessary duplication of animal studies and of publication bias and increase the quality of systematic reviews of animal experiments.

The assessment of the quality of systematic reviews, whether they are from an animal or human background, is still in its infancy. SR of AE would therefore greatly benefit from research into quality assessment systems, in order to get a clear overview of current quality and to point out further improvements.

Furthermore, future studies are needed to assess the development of quality of SR of AE over a larger period of time, as well as to look at procedures and practices to make SR of AE more feasible, for example initiatives like the CAMARADES research group.

Finally, using a step-by-step search guide when designing search strategies for SR of AE facilitates creating high quality SRs. Therefore the use of a guide, like the one developed by van Veggel *et al.* (in prep., see Appendix E) is highly encouraged by the author.

References

Altman, D. G. (2004) 'Improving Design and Analysis of Research: Lessons from clinical research', *Alternatives to Laboratory Animals,* **32**, (Supplement 2), 31-39.

Assendelft, W. J., van Tulder, M. W., Scholten, R. J. and Bouter, L. M. (1999) 'De Praktijjk van Systematische Reviews. II. Zoeken en selecteren van studies', *Nederlands Tijdschrift voor Geneeskunde,* **143**, (13), 656-661.

Bachmann, L. M., Estermann, P., Kronenberg, C. and ter Riet, G. (2003) 'Identifying Diagnostic Accuracy Studies in EMBASE', *Jounal of the Medical Library Association,* **91**, (3), 341-346.

Balls, M. (2009). *The Three Rs and the Humanity Criterion: An abridges version of "The Principles of Humane Experimental Technique".* Nottingham: FRAME.

Baumans, V. (2005) 'Science-based assessment of animal welfare: laboratory animals', *Revue Scientifique Et Technique,* **24**, (2), 503-513.

Bracken, M. B. (2009) 'Why animal studies are often poor predictors of human reactions to exposure', *Journal of the Royal Society of Medicine,* **102**, 120-122.

Chalmers, I. and Haynes, B. (1994) 'Reporting, updating, and correcting systematic reviews of the effects of health care', *British Medical Journal,* **309**, (6958), 862-865.

Chalmers, I. and Matthews, R. (2006) 'What are the implications of optimism bias in clinical research?', *The Lancet,* **367**, (9509), 449-450.

Chilov, M., Matsoukas, K., Ispahany, N., Allen, T. Y. and Lustbader, J. W. (2007) 'Using MeSH to search for alternatives to the use of animals in research', *Medical Reference Services Quarterly,* **26**, (3), 55-74.

Cochrane Collaboration. (2009). *Cochrane Handbook for Systematic Reviews of Interventions* 5.1.2. 5.1.2.

Darwin, C. (1859). *On the Origin of Species: By means of natural selection or the preervation of favoured races in the struggle for life.* London: Penguin Books.

Egger, M., Zellweger-Zahner, T., Schneider, M., Junker, C., Lengeler, C. and Antes, G. (1997) 'Language bias in randomised controlled trials published in English and German', *The Lancet,* **350**, (9074), 326-329.

Festing, M. F. W. (2004) 'The Choice of Animal Model and Reduction', *Alternatives to Laboratory Animals,* **32**, (Supplement 2), 59-64.

Gauthier, C. and Griffin, G. (2005) 'Using animals in research, testing and teaching', *Revue Scientifique Et Technique,* **24**, (2), 735-745.

Gerber, S., Tallon, D., Trelle, S., Schneider, M., Juni, P. and Egger, M. (2007) 'Bibliographic study showed improving methodology of meta-analyses published in leading journals 1993-2002', *Journal of Clinical Epidemiology,* **60**, (8), 773-780.

Green, S., Higgins, J. P. T., Alderson, P., Clarke, M., Mulrow, C. D. and Oxman, A. D. (2009). 'Introduction', In: J. P. T. Higgins and S. Green (Eds.), *Cochrane Handbook for Systematic Reviews of Interventions (Version 5.0.2)* (The Cochrane Collaboration,

Grégoire, G., Derderian, F. and Le Lorier, J. (1995) 'Selecting the language of the publications included in a meta-analysis: is there a Tower of Babel bias?', *Journal of Clinical Epidemiology,* **48**, (1), 159-163.

Guimaraes, C. A. (2009) 'Systematic review of animal research', *Acta Cirurgica Brasilia,* **24**, (1), 67-68.

Hackam, D. G. and Redelmeier, D. A. (2006) 'Translation of research evidence from animals to humans', *Journal of the American Medical Association,* **296**, (14), 1731-1732.

Hackam, D. G. (2007) 'Translating animal research into clinical benefit', *British Medical Journal,* **334**, (7586), 163-164.

Hart, L. A., Wood, M. W. and Weng, H. Y. (2005) 'Effective searching of the scientific literature for alternatives: search grids for appropriate databases', *Animal Welfare,* **14**, (4), 287-289.

Haynes, R. B., McKibbon, K. A., Wilczynski, N. L., Walter, S. D. and Werre, S. R. (2005) 'Optimal search strategies for retrieving scientifically strong studies of treatment from Medline: analytical survey', *British Medical Journal,* **330**, (7501), 1179.

Higgins, J. P. T. and Altman, D. G. (2009). 'Assessing Risk of Bias in Included Studies', In: J. P. T. Higgins and S. Green (Eds.), *Cochrane Handbook for Systematic Reviews of Interventions (Version 5.0.2)* (The Cochrane Collaboration,

Hooijmans, C. R., Leenaars, M. and Ritskes-Hoitinga, M. (2010a) 'A Gold Standard Publication Checklist to Improve the Quality of Animal Studies, to Fully Integrate the Three Rs, and to Make Systematic Reviews More Feasible', *Alternatives to Laboratory Animals,* **38**, 167-182.

Hooijmans, C. R., Tillema, A., Leenaars, M. and Ritskes-Hoitinga, M. (2010b) 'Enhancing search efficiency by means of a search filter for finding all studies on animal experimentation in PubMed', *Laboratory Animals,* **44**, (3), 170-175.

Horvath, A. R. and Pewsner, D. (2004) 'Systematic Reviews in Laboratory medicine: Principles, processes and practical considerations', *Clinica Chimica Acta,* **342**, 23-39.

Howard, B., Hudson, M. and Preziosi, R. (2009) 'More is Less: Reducing Animal Use by Raising Awareness of the Principles of Efficient Study Design and Analysis', *Alternatives to Laboratory Animals,* **37**, 33-42.

Jadad, A. R. and McQuay, H. J. (1996) 'Meta-analyses to evaluate analgesic interventions: a systematic qualitative review of their methodology', *Journal of Clinical Epidemiology*, **49**, (2), 235-243.

Jadad, A. R., Cook, D. J., Jones, A., Klassen, T. P., Tugwell, P., Moher, M. and Moher, D. (1998a) 'Methodology and reports of systematic reviews and meta-analyses: a comparison of Cochrane reviews with articles published in paper-based journals', *Journal of the American Medical Association*, **280**, (3), 278-280.

Jadad, A. R., Moher, D. and Klassen, T. P. (1998b) 'Guides for reading and interpreting systematic reviews: II. How did the authors find the studies and assess their quality?', *Archives of Pediatric and Adolescent Medicine*, **152**, (8), 812-817.

Jadad, A. R., Moher, M., Browman, G. P., Booker, L., Sigouin, C., Fuentes, M. and Stevens, R. (2000) 'Systematic reviews and meta-analyses on treatment of asthma: critical evaluation', *British Medical Journal*, **320**, (7234), 537-540.

Jüni, P., Altman, D. G. and Egger, M. (2001) 'Systematic reviews in health care: Assessing the quality of controlled clinical trials', *British Medical Journal*, **323**, (7303), 42-46.

Jüni, P., Holenstein, F., Sterne, J., Bartlett, C. and Egger, M. (2002) 'Direction and impact of language bias in meta-analyses of controlled trials: empirical study', *International Journal of Epidemiology*, **31**, (1), 115-123.

Khan, K. S. and Mignini, L. (2005) 'Surveying the literature from animal experiments: avoidance of bias is objective of systematic reviews, not meta-analysis', *British Medical Journal*, **331**, (7508), 110-111.

Kilkenny, C., Parsons, N., Kadyszewski, E., Festing, M. F., Cuthill, I. C., Fry, D., Hutton, J. and Altman, D. G. (2009) 'Survey of the quality of experimental design, statistical analysis and reporting of research using animals', *PLoS One*, **4**, (11), e7824.

Kilkenny, C., Browne, W. J., Cuthill, I. C., Emerson, M. and Altman, D. G. (2010) 'Improving bioscience research reporting: the ARRIVE guidelines for reporting animal research', *PLoS Biology*, **8**, (6), e1000412.

Knight, A. (2007) 'Systematic reviews of animal experiments demonstrate poor human clinical and toxicological utility', *Alternatives to Laboratory Animals*, **35**, (6), 641-659.

Knight, A. (2008) 'Reviewing existing knowledge prior to conducting animal studies', *Alternatives to Laboratory Animals*, **36**, (6), 709-712.

Krupski, T. L., Dahm, P., Fesperman, S. F. and Schardt, C. M. (2008) 'Users' Guide to the Urological Literature: How to perform a literature search', *The Journal of Urology*, **179**, 1264-1270.

Leenaars, M., Savenije, B., Nagtegaal, A., van der Vaart, L. and Ritskes-Hoitinga, M. (2009) 'Assessing the search for and implementation of the Three Rs: a survey among scientists', *Alternatives to Laboratory Animals*, **37**, (3), 297-303.

Lemon, R. and Dunnett, S. B. (2005) 'Surveying the literature from animal experiments', *British Medical Journal*, **330**, (7498), 977-978.

Lindl, T., Voelkel, M. and Kolar, R. (2005) '[Animal experiments in biomedical research. An evaluation of the clinical relevance of approved animal experimental projects]', *Alternativen zu Tierexperimenten*, **22**, (3), 143-151.

Macleod, M. R., Ebrahim, S. and Roberts, I. (2005) 'Surveying the literature from animal experiments: systematic review and meta-analysis are important contributions', *British Medical Journal*, **331**, (7508), 110.

Macleod, M. R., van der Worp, H. B., Sena, E. S., Howells, D. W., Dirnagl, U. and Donnan, G. A. (2008) 'Evidence for the efficacy of NXY-059 in experimental focal cerebral ischaemia is confounded by study quality', *Stroke*, **39**, (10), 2824-2829.

Macleod, M. R., Fisher, M., O'Collins, V., Sena, E. S., Dirnagl, U., Bath, P. M., Buchan, A., van der Worp, H. B., Traystman, R., Minematsu, K., Donnan, G. A. and Howells, D. W. (2009) 'Good laboratory practice: preventing introduction of bias at the bench', *Stroke*, **40**, (3), e50-52.

Mignini, L. E. and Khan, K. S. (2006) 'Methodological quality of systematic reviews of animal studies: a survey of reviews of basic research', *BMC Medical Research Methodology*, **6**, 10.

Moher, D., Fortin, P., Jadad, A. R., Juni, P., Klassen, T., Le Lorier, J., Liberati, A., Linde, K. and Penna, A. (1996) 'Completeness of reporting of trials published in languages other than English. implications for conduct and reporting of systematic reviews', *The Lancet*, **347**, (8998), 363-366.

Moher, D., Cook, D. J., Eastwood, S., Olkin, I., Rennie, D. and Stroup, D. F. (1999) 'Improving the quality of reports of meta-analyses of randomised controlled trials: the QUOROM statement. Quality of Reporting of Meta-analyses', *The Lancet*, **354**, (9193), 1896-1900.

Moher, D., Pham, B., Lawson, M. L. and Klassen, T. P. (2003) 'The inclusion of reports of randomised trials published in languages other than English in systematic reviews', *Health Technology Assessment*, **7**, (41), 1-90.

Moher, D., Tetzlaff, J., Tricco, A. C., Sampson, M. and Altman, D. G. (2007) 'Epidemiology and reporting characteristics of systematic reviews', *PLoS Medicine*, **4**, (3), e78.

Moher, D., Liberati, A., Tetzlaff, J. and Altman, D. G. (2009) 'Preferred reporting items for systematic reviews and meta-analyses: the PRISMA statement', *PLoS Medicine*, **6**, (7), e1000097.

Moja, L. P., Telaro, E., D'Amico, R., Moschetti, I., Coe, L. and Liberati, A. (2005) 'Assessment of methodological quality of primary studies by systematic reviews: results of the metaquality cross sectional study', *British Medical Journal,* **330**, (7499), 1053.

Montori, V. M., Devereaux, P. J., Adhikari, N. K., Burns, K. E., Eggert, C. H., Briel, M., Lacchetti, C., Leung, T. W., Darling, E., Bryant, D. M., Bucher, H. C., Schunemann, H. J., Meade, M. O., Cook, D. J., Erwin, P. J., Sood, A., Sood, R., Lo, B., Thompson, C. A., Zhou, Q., Mills, E. and Guyatt, G. H. (2005a) 'Randomized trials stopped early for benefit: a systematic review', *Journal of the American Medical Association,* **294**, (17), 2203-2209.

Montori, V. M., Wilczynski, N. L., Morgan, D. and Haynes, R. B. (2005b) 'Optimal search strategies for retrieving systematic reviews from Medline: analytical survey', *British Medical Journal,* **330**, (7482), 68.

O'Connor, A. M., Sargeant, J. M., Gardner, I. A., Dickson, J. S., Torrence, M. E., Dewey, C. E., Dohoo, I. R., Evans, R. B., Gray, J. T., Greiner, M., Keefe, G., Lefebvre, S. L., Morley, P. S., Ramirez, A., Sischo, W., Smith, D. R., Snedeker, K., Sofos, J., Ward, M. P. and Wills, R. (2010) 'The REFLECT statement: methods and processes of creating reporting guidelines for randomized controlled trials for livestock and food safety', *Preventive Veterinary Medicine,* **93**, (1), 11-18.

Offringa, M. and de Craen, A. J. (1999) 'De Praktijk van Systematische Reviews. I. Introductie', *Nederlands Tijdschrift voor Geneeskunde,* **143**, (13), 653-656.

Olivry, T. and Bizikova, P. (2010) 'A systematic review of the evidence of reduced allergenicity and clinical benefit of food hydrolysates in dogs with cutaneous adverse food reactions', *Veterinary Dermatology,* **21**, (1), 32-41.

Pai, M., McCulloch, M., Gorman, J. D., Pai, N., Enanoria, W., Kennedy, G., Tharyan, P. and Colford, J. M., Jr. (2004) 'Systematic reviews and meta-analyses: an illustrated, step-by-step guide', *National Medical Journal of India,* **17**, (2), 86-95.

Perel, P., Roberts, I., Sena, E., Wheble, P., Briscoe, C., Sandercock, P., Macleod, M., Mignini, L. E., Jayaram, P. and Khan, K. S. (2007) 'Comparison of treatment effects between animal experiments and clinical trials: systematic review', *British Medical Journal,* **334**, (7586), 197.

Peters, J. L., Sutton, A. J., Jones, D. R., Rushton, L. and Abrams, K. R. (2006) 'A systematic review of systematic reviews and meta-analyses of animal experiments with guidelines for reporting', *Journal of Environmental Science and Health, Part B,* **41**, (7), 1245-1258.

Phillips, C. J. C. (2005) 'Meta-analysis - a systematic and quantitative review of animal experiments to maximise the information derived', *Animal Welfare,* **14**, (4), 333-338.

Pound, P., Ebrahim, S., Sandercock, P., Bracken, M. B. and Roberts, I. (2004) 'Where is the evidence that animal research benefits humans?', *British Medical Journal*, **328**, (7438), 514-517.

Roberts, I., Kwan, I., Evans, P. and Haig, S. (2002) 'Does animal experimentation inform human healthcare? Observations from a systematic review of international animal experiments on fluid resuscitation', *British Medical Journal*, **324**, (7335), 474-476.

Sampson, M. and McGowan, J. (2006) 'Errors in search strategies were identified by type and frequency', *Journal of Clinical Epidemiology*, **59**, (10), 1057-1063.

Sampson, M., McGowan, J., Tetzlaff, J., Cogo, E. and Moher, D. (2008) 'No consensus exists on search reporting methods for systematic reviews', *Journal of Clinical Epidemiology*, **61**, (8), 748-754.

Savoie, I., Helmer, D., Green, C. J. and Kazanjian, A. (2003) 'Beyond Medline: reducing bias through extended systematic review search', *International Journal of Technology Assessment in Health Care*, **19**, (1), 168-178.

Schulz, K. F., Altman, D. G. and Moher, D. (2010) 'CONSORT 2010 Statement: Updated guidelines for reporting parallel group randomised trials', *Journal of Clinical Epidemiology*.

Sena, E., Wheble, P., Sandercock, P. and Macleod, M. (2007) 'Systematic review and meta-analysis of the efficacy of tirilazad in experimental stroke', *Stroke*, **38**, (2), 388-394.

Sena, E., van der Worp, H. B., Howells, D. W. and Macleod, M. (2010a) 'How can we improve the pre-clinical development of drugs for stroke', *Trends in Neurosciences*, **in press**.

Sena, E. S., van der Worp, H. B., Bath, P. M., Howells, D. W. and Macleod, M. R. (2010b) 'Publication bias in reports of animal stroke studies leads to major overstatement of efficacy', *PLoS Biology*, **8**, (3), e1000344.

Shea, B., Bouter, L. M., Grimshaw, J. M., Francis, D., Ortiz, Z., Wells, G. A., Tugwell, P. S. and Boers, M. (2006) 'Scope for improvement in the quality of reporting of systematic reviews. From the Cochrane Musculoskeletal Group', *Journal of Rheumatology*, **33**, (1), 9-15.

Shojania, K. G., Sampson, M., Ansari, M. T., Ji, J., Doucette, S. and Moher, D. (2007) 'How quickly do systematic reviews go out of date? A survival analysis', *Annals of Internal Medicine*, **147**, (4), 224-233.

Simera, I., Moher, D., Hirst, A., Hoey, J., Schulz, K. F. and Altman, D. G. (2010) 'Transparent and accurate reporting increases reliability, utility, and impact of your research: reporting guidelines and the EQUATOR Network', *BMC Medicine*, **8**, (1), 24.

Starr, M., Chalmers, I., Clarke, M. and Oxman, A. D. (2009) 'The origins, evolution, and future of The Cochrane Database of Systematic Reviews', *International Journal of Technology Assessment in Health Care*, **25 Suppl 1**, 182-195.

Tricco, A. C., Pham, B., Brehaut, J., Tetroe, J., Cappelli, M., Hopewell, S., Lavis, J. N., Berlin, J. A. and Moher, D. (2009) 'An international survey indicated that unpublished systematic reviews exist', *Journal of Clinical Epidemiology,* **62**, (6), 617-623 e615.

van der Worp, H. B., de Haan, P., Morrema, E. and Kalkman, C. J. (2005) 'Methodological quality of animal studies on neuroprotection in focal cerebral ischaemia', *Journal of Neurology,* **252**, 1108-1114.

van der Worp, H. B., Howells, D. W., Sena, E. S., Porrit, M. J., S., R., O'Collins, V. and Macleod, M. R. (2010) 'Can Animal Models of Disease Reliably Inform Human Studies?', *PLoS Medicine,* **7**, (3), e1000245.

van Zutphen, L. F. M. and Ohl, F. (2009). 'Inleiding', In: L. F. M. v. Zutphen, V. Baumans and F. Ohl (Eds.), *Handboek Proefdierkunde: Proefdieren, dierproeven, alternatieven en ethiek* (Vijfde druk edition). Maarssen: Elsevier Gezondheidszorg, 17-25

Wheble, P. C., Sena, E. S. and Macleod, M. R. (2008) 'A systematic review and meta-analysis of the efficacy of piracetam and piracetam-like compounds in experimental stroke', *Cerebrovascular Diseases,* **25**, (1-2), 5-11.

Wong, S. S., Wilczynski, N. L. and Haynes, R. B. (2006) 'Developing optimal search strategies for detecting clinically sound treatment studies in EMBASE', *Journal of the Medical Library Association,* **94**, (1), 41-47.

Wood, M. W. and Hart, L. A. (2007) 'Selecting appropriate animal models and strains: Making the best use of research, information and outreach', *Alternatives to Animal Testing and Experimentation,* **14**, (Special Issue), 303-306.

Woods, D. and Trewheellar, K. (1998) 'Medline and Embase complement each other in literature searches', *British Medical Journal,* **316**, (7138), 1166.

Yoshii, A., Plaut, D. A., McGraw, K. A., Anderson, M. J. and Wellik, K. E. (2009) 'Analysis of the reporting of search strategies in Cochrane systematic reviews', *Journal of the Medical Library Association,* **97**, (1), 21-29.

Appendix A: Animal Experiments filters for PubMed and EMBASE

PubMed

("animal experimentation"[MeSH Terms] OR "models, animal"[MeSH Terms] OR "invertebrates"[MeSH Terms] OR "Animals"[Mesh:noexp] OR "animal population groups"[MeSH Terms] OR "chordata"[MeSH Terms:noexp] OR "chordata, nonvertebrate"[MeSH Terms] OR "vertebrates"[MeSH Terms:noexp] OR "amphibians"[MeSH Terms] OR "birds"[MeSH Terms] OR "fishes"[MeSH Terms] OR "reptiles"[MeSH Terms] OR "mammals"[MeSH Terms:noexp] OR "primates"[MeSH Terms:noexp] OR "artiodactyla"[MeSH Terms] OR "carnivora"[MeSH Terms] OR "cetacea"[MeSH Terms] OR "chiroptera"[MeSH Terms] OR "elephants"[MeSH Terms] OR "hyraxes"[MeSH Terms] OR "insectivora"[MeSH Terms] OR "lagomorpha"[MeSH Terms] OR "marsupialia"[MeSH Terms] OR "monotremata"[MeSH Terms] OR "perissodactyla"[MeSH Terms] OR "rodentia"[MeSH Terms] OR "scandentia"[MeSH Terms] OR "sirenia"[MeSH Terms] OR "xenarthra"[MeSH Terms] OR "haplorhini"[MeSH Terms:noexp] OR "strepsirhini"[MeSH Terms] OR "platyrrhini"[MeSH Terms] OR "tarsii"[MeSH Terms] OR "catarrhini"[MeSH Terms:noexp] OR "cercopithecidae"[MeSH Terms] OR "hylobatidae"[MeSH Terms] OR "hominidae"[MeSH Terms:noexp] OR "gorilla gorilla"[MeSH Terms] OR "pan paniscus"[MeSH Terms] OR "pan troglodytes"[MeSH Terms] OR "pongo pygmaeus"[MeSH Terms]) OR ((animals[tiab] OR animal[tiab] OR mice[Tiab] OR mus[Tiab] OR mouse[Tiab] OR murine[Tiab] OR woodmouse[tiab] OR rats[Tiab] OR rat[Tiab] OR murinae[Tiab] OR muridae[Tiab] OR cottonrat[tiab] OR cottonrats[tiab] OR hamster[tiab] OR hamsters[tiab] OR cricetinae[tiab] OR rodentia[Tiab] OR rodent[Tiab] OR rodents[Tiab] OR pigs[Tiab] OR pig[Tiab] OR swine[tiab] OR swines[tiab] OR piglets[tiab] OR piglet[tiab] OR boar[tiab] OR boars[tiab] OR "sus scrofa"[tiab] OR ferrets[tiab] OR ferret[tiab] OR polecat[tiab] OR polecats[tiab] OR "mustela putorius"[tiab] OR "guinea pigs"[Tiab] OR "guinea pig"[Tiab] OR cavia[Tiab] OR callithrix[Tiab] OR marmoset[Tiab] OR marmosets[Tiab] OR cebuella[Tiab] OR hapale[Tiab] OR octodon[Tiab] OR chinchilla[Tiab] OR chinchillas[Tiab] OR gerbillinae[Tiab] OR gerbil[Tiab] OR gerbils[Tiab] OR jird[Tiab] OR jirds[Tiab] OR merione[Tiab] OR meriones[Tiab] OR rabbits[Tiab] OR rabbit[Tiab] OR hares[Tiab] OR hare[Tiab] OR diptera[Tiab] OR flies[Tiab] OR fly[Tiab] OR dipteral[Tiab] OR drosphila[Tiab] OR drosophilidae[Tiab] OR cats[Tiab] OR cat[Tiab] OR carus[Tiab] OR felis[Tiab] OR nematoda[Tiab] OR nematode[Tiab] OR nematoda[Tiab] OR nematode[Tiab] OR nematodes[Tiab] OR sipunculida[Tiab] OR dogs[Tiab] OR dog[Tiab] OR canine[Tiab] OR canines[Tiab] OR canis[Tiab] OR sheep[Tiab] OR sheeps[Tiab] OR mouflon[Tiab] OR mouflons[Tiab] OR ovis[Tiab] OR goats[Tiab] OR goat[Tiab] OR capra[Tiab] OR capras[Tiab] OR rupicapra[Tiab] OR chamois[Tiab] OR haplorhini[Tiab] OR monkey[Tiab] OR monkeys[Tiab] OR anthropoidea[Tiab] OR anthropoids[Tiab] OR saguinus[Tiab] OR tamarin[Tiab] OR tamarins[Tiab] OR leontopithecus[Tiab] OR hominidae[Tiab] OR ape[Tiab] OR apes[Tiab] OR pan[Tiab] OR paniscus[Tiab] OR "pan paniscus"[Tiab] OR bonobo[Tiab] OR bonobos[Tiab] OR troglodytes[Tiab] OR "pan troglodytes"[Tiab] OR gibbon[Tiab] OR gibbons[Tiab] OR siamang[Tiab] OR siamangs[Tiab] OR nomascus[Tiab] OR symphalangus[Tiab] OR chimpanzee[Tiab] OR chimpanzees[Tiab] OR prosimians[Tiab] OR "bush baby"[Tiab] OR prosimian[Tiab] OR bush babies[Tiab] OR galagos[Tiab] OR galago[Tiab] OR pongidae[Tiab] OR gorilla[Tiab] OR gorillas[Tiab] OR pongo[Tiab] OR pygmaeus[Tiab] OR "pongo pygmaeus"[Tiab] OR orangutans[Tiab] OR pygmaeus[Tiab] OR lemur[Tiab] OR lemurs[Tiab] OR lemuridae[Tiab] OR horse[Tiab] OR horses[Tiab] OR pongo[Tiab] OR equus[Tiab] OR cow[Tiab] OR calf[Tiab] OR bull[Tiab] OR chicken[Tiab] OR chickens[Tiab] OR gallus[Tiab] OR quail[Tiab] OR bird[Tiab] OR birds[Tiab] OR quails[Tiab] OR poultry[Tiab] OR poultries[Tiab] OR fowl[Tiab] OR fowls[Tiab] OR reptile[Tiab] OR reptilia[Tiab] OR reptiles[Tiab] OR snakes[Tiab] OR snake[Tiab] OR lizard[Tiab] OR lizards[Tiab] OR alligator[Tiab] OR alligators[Tiab] OR crocodile[Tiab] OR crocodiles[Tiab] OR turtle[Tiab] OR turtles[Tiab] OR amphibian[Tiab] OR amphibians[Tiab] OR amphibia[Tiab] OR frog[Tiab] OR frogs[Tiab] OR bombina[Tiab] OR salientia[Tiab] OR toad[Tiab] OR toads[Tiab] OR "epidalea calamita"[Tiab] OR salamander[Tiab] OR salamanders[Tiab] OR eel[Tiab] OR eels[Tiab] OR fish[Tiab] OR fishes[Tiab] OR pisces[Tiab] OR catfish[Tiab] OR catfishes[Tiab] OR siluriformes[Tiab] OR arius[Tiab] OR heteropneustes[Tiab] OR sheatfish[Tiab] OR perch[Tiab] OR perches[Tiab] OR percidae[Tiab] OR perca[Tiab] OR trout[Tiab] OR trouts[Tiab] OR char[Tiab] OR chars[Tiab] OR salvelinus[Tiab] OR "fathead minnow"[Tiab] OR minnow[Tiab] OR cyprinidae[Tiab] OR carps[Tiab] OR carp[Tiab] OR zebrafish[Tiab] OR zebrafishes[Tiab] OR goldfish[Tiab] OR goldfishes[Tiab] OR guppy[Tiab] OR guppies[Tiab] OR chub[Tiab] OR chubs[Tiab] OR tinca[Tiab] OR barbels[Tiab] OR barbus[Tiab] OR pimephales[Tiab] OR promelas[Tiab] OR "poecilia reticulata"[Tiab] OR mullet[Tiab] OR mullets[Tiab] OR seahorse[Tiab] OR seahorses[Tiab] OR mugil curema[Tiab] OR atlantic cod[Tiab] OR shark[Tiab] OR sharks[Tiab] OR catshark[Tiab] OR anguilla[Tiab] OR salmonid[Tiab] OR salmonids[Tiab] OR whitefish[Tiab] OR whitefishes[Tiab] OR salmon[Tiab] OR salmons[Tiab] OR sole[Tiab] OR solea[Tiab] OR "sea lamprey"[Tiab] OR lamprey[Tiab] OR lampreys[Tiab] OR pumpkinseed[Tiab] OR sunfish[Tiab] OR sunfishes[Tiab] OR tilapia[Tiab] OR tilapias[Tiab] OR turbot[Tiab] OR turbots[Tiab] OR flatfish[Tiab] OR flatfishes[Tiab] OR sciuridae[Tiab] OR squirrel[Tiab] OR squirrels[Tiab] OR chipmunk[Tiab] OR chipmunks[Tiab] OR suslik[Tiab] OR susliks[Tiab] OR vole[Tiab] OR voles[Tiab] OR lemming[Tiab] OR lemmings[Tiab] OR muskrat[Tiab] OR muskrats[Tiab] OR lemmus[Tiab] OR otter[Tiab] OR otters[Tiab] OR marten[Tiab] OR martens[Tiab] OR martes[Tiab] OR weasel[Tiab] OR badger[Tiab] OR badgers[Tiab] OR ermine[Tiab] OR mink[Tiab] OR minks[Tiab] OR sable[Tiab] OR sables[Tiab] OR gulo[Tiab] OR gulos[Tiab] OR wolverine[Tiab] OR wolverines[Tiab] OR minks[Tiab] OR mustela[Tiab] OR llama[Tiab] OR llamas[Tiab] OR alpaca[Tiab] OR alpacas[Tiab] OR camelid[Tiab] OR camelids[Tiab] OR guanaco[Tiab] OR guanacos[Tiab] OR chiroptera[Tiab] OR chiropteras[Tiab] OR bat[Tiab] OR bats[Tiab] OR fox[Tiab] OR foxes[Tiab] OR iguana[Tiab] OR iguanas[Tiab] OR xenopus laevis[Tiab] OR parakeet[Tiab] OR parakeets[Tiab] OR parrot[Tiab] OR parrots[Tiab] OR donkey[Tiab] OR donkeys[Tiab] OR mule[Tiab] OR mules[Tiab] OR zebra[Tiab] OR zebras[Tiab] OR shrew[Tiab] OR shrews[Tiab] OR bison[Tiab] OR bisons[Tiab] OR buffalo[Tiab] OR buffaloes[Tiab] OR deer[Tiab] OR deers[Tiab] OR bear[Tiab] OR bears[Tiab] OR panda[Tiab] OR pandas[Tiab] OR "wild hog"[Tiab] OR "wild boar"[Tiab] OR fitchew[Tiab] OR fitch[Tiab] OR beaver[Tiab] OR beavers[Tiab] OR jerboa[Tiab] OR jerboas[Tiab] OR capybara[Tiab] OR capybaras[Tiab]) NOT medline[subset])

EMBASE

Exp animal experimentation/ OR exp animal model/ OR exp Invertebrate/ OR exp animal/ OR exp chordata/ OR exp experimental animal/ OR exp transgenic animal/ OR exp adult animal/ OR exp female animal/ OR exp game animal/ OR exp juvenile animal/ OR exp male animal/ OR exp domestic animal/ OR exp feral animal/ OR exp wild animal/ OR exp contaminated animal/ OR exp contaminated fish/ OR exp contaminated mussel/ OR exp contaminated shellfish/ OR exp poisonous animal/ OR exp poisonous caterpillar/ OR exp poisonous frog/ OR exp poisonous insect/ OR exp poisonous jellyfish/ OR exp poisonous scorpion/ OR exp poisonous snake/ OR exp poisonous spider/ OR exp toxic fish/ OR exp toxic octopus/ OR exp toxic sea urchin/ OR exp toxic slug/ OR exp toxic snail/ OR exp amphibia/ OR exp bird/ OR exp fish/ OR exp reptile/ OR exp hyrax/ OR exp marsupial/ OR exp monotremate/ OR exp scandentia/ OR exp bat/ OR exp carnivora/ OR exp cetacea/ OR exp edentata/ OR exp elephant/ OR exp insectivora/ OR exp lagomorph/ OR exp rodent/ OR exp sirenia/ OR exp ungulate/ OR exp prosimian/ OR exp aotus/ OR exp ape/ OR exp atelidae/ OR exp catarrhini/ OR exp cebidae/ OR exp cercocebus/ OR exp cercopithecidae/ OR exp colobinae/ OR exp haplorhini/ OR exp leontopithecus/ OR exp macaca/ OR exp monkey/ OR exp pitheciidae/ OR exp platyrrhini/ OR exp tarsier/ OR animals.ti,ab. OR animal.ti,ab. OR mice.ti,ab. OR mus.ti,ab. OR mouse.ti,ab. OR murine.ti,ab. OR woodmouse.ti,ab. OR rats.ti,ab. OR rat.ti,ab. OR murinae.ti,ab. OR muridae.ti,ab. OR cottonrat.ti,ab. OR cottonrats.ti,ab. OR hamster.ti,ab. OR hamsters.ti,ab. OR cricetinae.ti,ab. OR rodentia.ti,ab. OR rodent.ti,ab. OR rodents.ti,ab. OR pigs.ti,ab. OR pig.ti,ab. OR swine.ti,ab. OR swines.ti,ab. OR piglets.ti,ab. OR piglet.ti,ab. OR boar.ti,ab. OR boars.ti,ab. OR "sus scrofa".ti,ab. OR ferrets.ti,ab. OR ferret.ti,ab. OR polecat.ti,ab. OR polecats.ti,ab. OR "mustela putorius".ti,ab. OR "guinea pigs".ti,ab. OR "guinea pig".ti,ab. OR cavia.ti,ab. OR callithrix.ti,ab. OR marmoset.ti,ab. OR marmosets.ti,ab. OR cebuella.ti,ab. OR hapale.ti,ab. OR octodon.ti,ab. OR chinchilla.ti,ab. OR chinchillas.ti,ab. OR gerbillinae.ti,ab. OR gerbil.ti,ab. OR gerbils.ti,ab. OR jird.ti,ab. OR jirds.ti,ab. OR merione.ti,ab. OR meriones.ti,ab. OR rabbits.ti,ab. OR rabbit.ti,ab. OR hares.ti,ab. OR hare.ti,ab. OR diptera.ti,ab. OR flies.ti,ab. OR fly.ti,ab. OR dipteral.ti,ab. OR drosphila.ti,ab. OR drosophilidae.ti,ab. OR cats.ti,ab. OR cat.ti,ab. OR carus.ti,ab. OR felis.ti,ab. OR nematoda.ti,ab. OR nematode.ti,ab. OR nematoda.ti,ab. OR nematode.ti,ab. OR nematodes.ti,ab. OR sipunculida.ti,ab. OR dogs.ti,ab. OR dog.ti,ab. OR canine.ti,ab. OR canines.ti,ab. OR canis.ti,ab. OR sheep.ti,ab. OR sheeps.ti,ab. OR mouflon.ti,ab. OR mouflons.ti,ab. OR ovis.ti,ab. OR goats.ti,ab. OR goat.ti,ab. OR capra.ti,ab. OR capras.ti,ab. OR rupicapra.ti,ab. OR chamois.ti,ab. OR haplorhini.ti,ab. OR monkey.ti,ab. OR monkeys.ti,ab. OR anthropoidea.ti,ab. OR anthropoids.ti,ab. OR saguinus.ti,ab. OR tamarin.ti,ab. OR tamarins.ti,ab. OR leontopithecus.ti,ab. OR hominidae.ti,ab. OR ape.ti,ab. OR apes.ti,ab. OR pan.ti,ab. OR paniscus.ti,ab. OR "pan paniscus".ti,ab. OR bonobo.ti,ab. OR bonobos.ti,ab. OR troglodytes.ti,ab. OR "pan troglodytes".ti,ab. OR gibbon.ti,ab. OR gibbons.ti,ab. OR siamang.ti,ab. OR siamangs.ti,ab. OR nomascus.ti,ab. OR symphalangus.ti,ab. OR chimpanzee.ti,ab. OR chimpanzees.ti,ab. OR prosimians.ti,ab. OR "bush baby".ti,ab. OR prosimian.ti,ab. OR bush babies.ti,ab. OR galagos.ti,ab. OR galago.ti,ab. OR pongidae.ti,ab. OR gorilla.ti,ab. OR gorillas.ti,ab. OR pongo.ti,ab. OR pygmaeus.ti,ab. OR "pongo pygmaeus".ti,ab. OR orangutans.ti,ab. OR pygmaeus.ti,ab. OR lemur.ti,ab. OR lemurs.ti,ab. OR lemuridae.ti,ab. OR horse.ti,ab. OR horses.ti,ab. OR pongo.ti,ab. OR equus.ti,ab. OR cow.ti,ab. OR calf.ti,ab. OR bull.ti,ab. OR chicken.ti,ab. OR chickens.ti,ab. OR gallus.ti,ab. OR quail.ti,ab. OR bird.ti,ab. OR birds.ti,ab. OR quails.ti,ab. OR poultry.ti,ab. OR poultries.ti,ab. OR fowl.ti,ab. OR fowls.ti,ab. OR reptile.ti,ab. OR reptilia.ti,ab. OR reptiles.ti,ab. OR snakes.ti,ab. OR snake.ti,ab. OR lizard.ti,ab. OR lizards.ti,ab. OR alligator.ti,ab. OR alligators.ti,ab. OR crocodile.ti,ab. OR crocodiles.ti,ab. OR turtle.ti,ab. OR turtles.ti,ab. OR amphibian.ti,ab. OR amphibians.ti,ab. OR amphibia.ti,ab. OR frog.ti,ab. OR frogs.ti,ab. OR bombina.ti,ab. OR salientia.ti,ab. OR toad.ti,ab. OR toads.ti,ab. OR "epidalea calamita".ti,ab. OR salamander.ti,ab. OR salamanders.ti,ab. OR eel.ti,ab. OR eels.ti,ab. OR fish.ti,ab. OR fishes.ti,ab. OR pisces.ti,ab. OR catfish.ti,ab. OR catfishes.ti,ab. OR siluriformes.ti,ab. OR arius.ti,ab. OR heteropneustes.ti,ab. OR sheatfish.ti,ab. OR perch.ti,ab. OR perches.ti,ab. OR percidae.ti,ab. OR perca.ti,ab. OR trout.ti,ab. OR trouts.ti,ab. OR char.ti,ab. OR chars.ti,ab. OR salvelinus.ti,ab. OR "fathead minnow".ti,ab. OR minnow.ti,ab. OR cyprinidae.ti,ab. OR carps.ti,ab. OR carp.ti,ab. OR zebrafish.ti,ab. OR zebrafishes.ti,ab. OR goldfish.ti,ab. OR goldfishes.ti,ab. OR guppy.ti,ab. OR guppies.ti,ab. OR chub.ti,ab. OR chubs.ti,ab. OR tinca.ti,ab. OR barbels.ti,ab. OR barbus.ti,ab. OR pimephales.ti,ab. OR promelas.ti,ab. OR "poecilia reticulata".ti,ab. OR mullet.ti,ab. OR mullets.ti,ab. OR seahorse.ti,ab. OR seahorses.ti,ab. OR mugil curema.ti,ab. OR atlantic cod.ti,ab. OR shark.ti,ab. OR sharks.ti,ab. OR catshark.ti,ab. OR anguilla.ti,ab. OR salmonid.ti,ab. OR salmonids.ti,ab. OR whitefish.ti,ab. OR whitefishes.ti,ab. OR salmon.ti,ab. OR salmons.ti,ab. OR sole.ti,ab. OR solea.ti,ab. OR "sea lamprey".ti,ab. OR lamprey.ti,ab. OR lampreys.ti,ab. OR pumpkinseed.ti,ab. OR sunfish.ti,ab. OR sunfishes.ti,ab. OR tilapia.ti,ab. OR tilapias.ti,ab. OR turbot.ti,ab. OR turbots.ti,ab. OR flatfish.ti,ab. OR flatfishes.ti,ab. OR sciuridae.ti,ab. OR squirrel.ti,ab. OR squirrels.ti,ab. OR chipmunk.ti,ab. OR chipmunks.ti,ab. OR suslik.ti,ab. OR susliks.ti,ab. OR vole.ti,ab. OR voles.ti,ab. OR lemming.ti,ab. OR lemmings.ti,ab. OR muskrat.ti,ab. OR muskrats.ti,ab. OR lemmus.ti,ab. OR otter.ti,ab. OR otters.ti,ab. OR marten.ti,ab. OR martens.ti,ab. OR martes.ti,ab. OR weasel.ti,ab. OR badger.ti,ab. OR badgers.ti,ab. OR ermine.ti,ab. OR mink.ti,ab. OR minks.ti,ab. OR sable.ti,ab. OR sables.ti,ab. OR gulo.ti,ab. OR gulos.ti,ab. OR wolverine.ti,ab. OR wolverines.ti,ab. OR minks.ti,ab. OR mustela.ti,ab. OR llama.ti,ab. OR llamas.ti,ab. OR alpaca.ti,ab. OR alpacas.ti,ab. OR camelid.ti,ab. OR camelids.ti,ab. OR guanaco.ti,ab. OR guanacos.ti,ab. OR chiroptera.ti,ab. OR chiropteras.ti,ab. OR bat.ti,ab. OR bats.ti,ab. OR fox.ti,ab. OR foxes.ti,ab. OR iguana.ti,ab. OR iguanas.ti,ab. OR xenopus laevis.ti,ab. OR parakeet.ti,ab. OR parakeets.ti,ab. OR parrot.ti,ab. OR parrots.ti,ab. OR donkey.ti,ab. OR donkeys.ti,ab. OR mule.ti,ab. OR mules.ti,ab. OR zebra.ti,ab. OR zebras.ti,ab. OR shrew.ti,ab. OR shrews.ti,ab. OR bison.ti,ab. OR bisons.ti,ab. OR buffalo.ti,ab. OR buffaloes.ti,ab. OR deer.ti,ab. OR deers.ti,ab. OR bear.ti,ab. OR bears.ti,ab. OR panda.ti,ab. OR pandas.ti,ab. OR "wild hog".ti,ab. OR "wild boar".ti,ab. OR fitchew.ti,ab. OR fitch.ti,ab. OR beaver.ti,ab. OR beavers.ti,ab. OR jerboa.ti,ab. OR jerboas.ti,ab. OR capybara.ti,ab. OR capybaras.ti,ab.

Appendix B: Systematic Review filters for PubMed and EMBASE

PubMed

(systematic review [ti] OR meta-analysis [pt] OR meta-analysis [ti] OR systematic literature review [ti] OR (systematic review [tiab] AND review [pt]) OR consensus development conference [pt] OR practice guideline [pt] OR cochrane database syst rev [ta] OR acp journal club [ta] OR health technol assess [ta] OR evid rep technol assess summ [ta]) OR ((evidence based[ti] OR evidence-based medicine [mh] OR best practice* [ti] OR evidence synthesis [tiab])AND (review [pt] OR diseases category[mh] OR behavior and behavior mechanisms [mh] OR therapeutics [mh] OR evaluation studies[pt] OR validation studies[pt] OR guideline [pt])) OR ((systematic [tw] OR systematically [tw] OR critical [tiab] OR (study selection [tw]) OR (predetermined [tw] OR inclusion [tw] AND criteri* [tw]) OR exclusion criteri* [tw] OR main outcome measures [tw] OR standard of care [tw] OR standards of care [tw]) AND (survey [tiab] OR surveys [tiab] OR overview* [tw] OR review [tiab] OR reviews [tiab] OR search* [tw] OR handsearch [tw] OR analysis [tiab] OR critique [tiab] OR appraisal [tw] OR (reduction [tw]AND (risk [mh] OR risk [tw]) AND (death OR recurrence))) AND (literature [tiab] OR articles [tiab] OR publications [tiab] OR publication [tiab] OR bibliography [tiab] OR bibliographies [tiab] OR published [tiab] OR unpublished [tw] OR citation [tw] OR citations [tw] OR database [tiab] OR internet [tiab] OR textbooks [tiab] OR references [tw] OR scales [tw] OR papers [tw] OR datasets [tw] OR trials [tiab] OR meta-analy* [tw] OR (clinical [tiab] AND studies [tiab]) OR treatment outcome [mh] OR treatment outcome [tw])) NOT (letter [pt] OR newspaper article [pt] OR comment [pt])

EMBASE

(systematic review.ti. OR meta-analysis.pt. OR meta-analysis.ti. OR systematic literature review.ti. OR (systematic review.ti,ab. AND review.pt.) OR consensus development conference.pt. OR practice guideline.pt. OR cochrane database syst rev.ja. OR acp journal club.ja. OR health technol assess.ja. OR evid rep technol assess summ.ja.) OR ((evidence based.ti. OR evidence-based medicine/ OR best practice*.ti. OR evidence synthesis.ti,ab.) AND (review.pt. OR diseases category/ OR exp behavior and behavior mechanisms/ OR exp therapeutics/ OR evaluation studies.pt. OR validation studies.pt. OR guideline.pt.)) OR ((systematic.tw. OR systematically.tw. OR critical.ti,ab. OR (study selection.tw.) OR (predetermined.tw. OR inclusion.tw. AND criteri*.tw.) OR exclusion criteri*.tw. OR main outcome measures.tw. OR standard of care.tw. OR standards of care.tw.) AND (survey.ti,ab. OR surveys.ti,ab. OR overview*.tw. OR review.ti,ab. OR reviews.ti,ab. OR search*.tw. OR handsearch.tw. OR analysis.ti,ab. OR critique.ti,ab. OR appraisal.tw. OR (reduction.tw.AND (exp risk/ OR risk.tw.) AND (death OR recurrence))) AND (literature.ti,ab. OR articles.ti,ab.

OR publications.ti,ab. OR publication.ti,ab. OR bibliography.ti,ab. OR bibliographies.ti,ab. OR published.ti,ab. OR unpublished.tw. OR citation.tw. OR citations.tw. OR database.ti,ab. OR internet.ti,ab. OR textbooks.ti,ab. OR references.tw. OR scales.tw. OR papers.tw. OR datasets.tw. OR trials.ti,ab. OR meta-analy*.tw. OR (clinical.ti,ab. AND studies.ti,ab.) OR exp treatment outcome/ OR treatment outcome.tw.)) NOT (letter.pt. OR newspaper article.pt. OR comment.pt.)

Appendix C: List of SR of AE included in this study

Ahern, B. J., Parvizi, J., Boston, R. and Schaer, T. P. (2009) 'Preclinical animal models in single site cartilage defect testing: a systematic review', *Osteoarthritis and Cartilage,* **17**, (6), 705-713.

Akhtar, A. Z., Pippin, J. J. and Sandusky, C. B. (2009) 'Animal studies in spinal cord injury: A systematic review of methylprednisolone', *Alternatives to Laboratory Animals,* **37**, (1), 43-62.

Amarasingh, S., Macleod, M. R. and Whittle, I. R. (2009) 'What is the translational efficacy of chemotherapeutic drug research in neuro-oncology? A systematic review and meta-analysis of the efficacy of BCNU and CCNU in animal models of glioma', *Jounal of Neuro-Oncology,* **91**, (2), 117-125.

Aragon, C. L., Hofmeister, E. H. and Budsberg, S. C. (2007) 'Systematic review of clinical trials of treatments for osteoarthritis in dogs', *Journal of the American Veterinary Medical Association,* **230**, (4), 514-521.

Bailey, E. L., McCulloch, J., Sudlow, C. and Wardlaw, J. M. (2009) 'Potential animal models of lacunar stroke: a systematic review', *Stroke,* **40**, (6), e451-458.

Banwell, V., Sena, E. S. and Macleod, M. R. (2009) 'Systematic review and stratified meta-analysis of the efficacy of interleukin-1 receptor antagonist in animal models of stroke', *Jounal of Stroke Cerebrovascular Diseases,* **18**, (4), 269-276.

Begg, D. J. and Whittington, R. J. (2008) 'Experimental animal infection models for Johne's disease, an infectious enteropathy caused by Mycobacterium avium subsp. paratuberculosis', *Veterinary Journal,* **176**, (2), 129-145.

Benatar, M. (2007) 'Lost in translation: treatment trials in the SOD1 mouse and in human ALS', *Neurobiology of Disease,* **26**, (1), 1-13.

Booker, C. S. and Mann, J. I. (2008) 'Trans fatty acids and cardiovascular health: translation of the evidence base', *Nutrition, Metabolism & Cardiovascular Diseases,* **18**, (6), 448-456.

Bouzeghrane, F., Naggara, O., Kallmes, D. F., Berenstein, A. and Raymond, J. (2010) 'In vivo experimental intracranial aneurysm models: A systematic review', *American Journal of Neuroradiology,* **31**, (3), 418-423.

Carrillo, M., Ricci, L. A., Coppersmith, G. A. and Melloni Jr, R. H. (2009) 'The effect of increased serotonergic neurotransmission on aggression: A critical meta-analytical review of preclinical studies', *Psychopharmacology,* **205**, (3), 349-368.

Corpet, D. E. and Pierre, F. (2005) 'How good are rodent models of carcinogenesis in predicting efficacy in humans? A systematic review and meta-analysis of colon chemoprevention in rats, mice and men', *European Journal of Cancer,* **41**, (13), 1911-1922.

Devine, J. M. and Zafonte, R. D. (2009) 'Physical exercise and cognitive recovery in acquired brain injury: a review of the literature', *Physical Medicine and Rehabilitation,* **1**, (6), 560-575.

Djasim, U. M., Wolvius, E. B., van Neck, J. W., Weinans, H. and van der Wal, K. G. H. (2007) 'Recommendations for optimal distraction protocols for various animal models on the basis of a systematic review of the literature', *International Journal of Oral and Maxillofacial Surgery,* **36**, (10), 877-883.

England, T. J., Gibson, C. L. and Bath, P. M. W. (2009) 'Granulocyte-colony stimulating factor in experimental stroke and its effects on infarct size and functional outcome: A systematic review', *Brain Research Reviews,* **62**, (1), 71-82.

Fountoulakis, K. N., Vieta, E., Bouras, C., Notaridis, G., Giannakopoulos, P., Kaprinis, G. and Akiskal, H. (2008) 'A systematic review of existing data on long-term lithium therapy: Neuroprotective or neurotoxic?', *International Journal of Neuropsychopharmacology,* **11**, (2), 269-287.

Frampton, G. K., Jansch, S., Scott-Fordsmand, J. J., Rombke, J. and Van den Brink, P. J. (2006) 'Effects of pesticides on soil invertebrates in laboratory studies: a review and analysis using species sensitivity distributions', *Enivironmental Toxicology and Chemistry,* **25**, (9), 2480-2489.

Gibson, C. L., Gray, L. J., Bath, P. M. W. and Murphy, S. P. (2008) 'Progesterone for the treatment of experimental brain injury; a systematic review', *Brain,* **131**, (2), 318-328.

Gibson, C. L., Gray, L. J., Murphy, S. P. and Bath, P. M. (2006) 'Estrogens and experimental ischemic stroke: a systematic review', *Jounal of Cerebral Blood Flow & Metabolism,* **26**, (9), 1103-1113.

Gielkens, P. F., Bos, R. R., Raghoebar, G. M. and Stegenga, B. (2007) 'Is there evidence that barrier membranes prevent bone resorption in autologous bone grafts during the healing period? A systematic review', *International Jounal of Oral and Maxillofacial Implants,* **22**, (3), 390-398.

Gonzalez-Reyes, R. E., Gutierrez-Alvarez, A. M. and Moreno, C. B. (2007) 'Manganese and epilepsy: A systematic review of the literature', *Brain Research Reviews,* **53**, (2), 332-336.

Hainsworth, A. H. and Markus, H. S. (2008) 'Do in vivo experimental models reflect human cerebral small vessel disease? A systematic review', *Jounal of Cerebral Blood Flow & Metabolism,* **28**, (12), 1877-1891.

Hasani-Ranjbar, S., Larijani, B. and Abdollah, M. (2008) 'A systematic review of iranian medicinal plants useful in diabetes mellitus', *Archives of Medical Science,* **4**, (3), 285-292.

Hasani-Ranjbar, S., Nayebi, N., Larijani, B. and Abdollahi, M. (2009) 'A systematic review of the efficacy and safety of herbal medicines used in the treatment of obesity', *World Journal of Gastroenterology,* **15**, (25), 3073-3085.

Honiden, S. and Gong, M. N. (2009) 'Diabetes, insulin, and development of acute lung injury', *Critical Care Medicine,* **37**, (8), 2455-2464.

Iovino, A. and Scheen, A. J. (2010) '[Modulation of tissue exposure to cortisol, new perspective for reducing the metabolic risk associated with obesity]', *Revue Medicale de Liege,* **65**, (3), 140-146.

Jamaty, C., Bailey, B., Larocque, A., Notebaert, E., Sanogo, K. and Chauny, J. M. (2010) 'Lipid emulsions in the treatment of acute poisoning: A systematic review of human and animal studies', *Clinical Toxicology,* **48**, (1), 1-27.

Jerndal, M., Forsberg, K., Sena, E. S., MacLeod, M. R., O'Collins, V. E., Linden, T., Nilsson, M. and Howells, D. W. (2010) 'A systematic review and meta-analysis of erythropoietin in experimental stroke', *Journal of Cerebral Blood Flow and Metabolism,* **30**, (5), 961-968.

Jonasson, Z. (2005) 'Meta-analysis of sex differences in rodent models of learning and memory: A review of behavioral and biological data', *Neuroscience and Biobehavioral Reviews,* **28**, (8), 811-825.

Jung, R. E., Thoma, D. S. and Hammerle, C. H. (2008) 'Assessment of the potential of growth factors for localized alveolar ridge augmentation: a systematic review', *Journal of Clinical Periodontology,* **35**, (8 Suppl), 255-281.

Kastner, S. B. (2006) 'A2-agonists in sheep: a review', *Veterinary Anaesthesia and Analgesia,* **33**, (2), 79-96.

Kelly, F. E. and Nolan, J. P. (2010) 'The effects of mild induced hypothermia on the myocardium: A systematic review', *Anaesthesia,* **65**, (5), 505-515.

Ker, K., Perel, P. and Blackhall, K. (2009) 'Beta-2 receptor antagonists for traumatic brain injury: A systematic review of controlled trials in animal models', *CNS Neuroscience and Therapeutics,* **15**, (1), 52-64.

Kienle, G. S., Glockmann, A., Schink, M. and Kiene, H. (2009) 'Viscum album L. extracts in breast and gynaecological cancers: a systematic review of clinical and preclinical research', *Journal of Experimental and Clinical Cancer Research,* **28**, 79.

Lindner, T., Cockbain, A. J., El Masry, M. A., Katonis, P., Tsiridis, E., Schizas, C. and Tsiridis, E. (2008) 'The effect of anticoagulant pharmacotherapy on fracture healing', *Expert Opinion on Pharmacotherapy,* **9**, (7), 1169-1187.

Litle, M. P. and Lambert, B. E. (2008) 'Systematic review of experimental studies on the relative biological effectiveness of tritium', *Radiation and Environmental Biophysics,* **47**, (1), 71-93.

Macleod, M. R., O'Collins, T., Horky, L. L., Howells, D. W. and Donnan, G. A. (2005) 'Systematic review and metaanalysis of the efficacy of FK506 in experimental stroke', *Journal of Cerebral Blood Flow and Metabolism,* **25**, (6), 713-721.

Macleod, M. R., O'Collins, T., Horky, L. L., Howells, D. W. and Donnan, G. A. (2005) 'Systematic review and meta-analysis of the efficacy of melatonin in experimental stroke', *Journal of Pineal Research,* **38**, (1), 35-41.

MacRae, C. A. (2010) 'Cardiac arrhythmia: In vivo screening in the zebrafish to overcome complexity in drug discovery', *Expert Opinion on Drug Discovery,* **5**, (7), 619-632.

Matthan, N. R., Jordan, H., Chung, M., Lichtenstein, A. H., Lathrop, D. A. and Lau, J. (2005) 'A systematic review and meta-analysis of the impact of omega-3 fatty acids on selected arrhythmia outcomes in animal models', *Metabolism,* **54**, (12), 1557-1565.

Minnerup, J., Heidrich, J., Rogalewski, A., Schabitz, W. R. and Wellmann, J. (2009) 'The efficacy of erythropoietin and its analogues in animal stroke models: a meta-analysis', *Stroke,* **40**, (9), 3113-3120.

Miyairi, I., Ramsey, K. H. and Patton, D. L. (2010) 'Duration of untreated chlamydial genital infection and factors associated with clearance: review of animal studies', *The Journal of infectious diseases,* **2**, (pp S96-103).

Miyairi, I., Ramsey, K. H. and Patton, D. L. (2010) 'Duration of untreated chlamydial genital infection and factors associated with clearance: review of animal studies', *The Journal of infectious diseases,* **2**, (pp S96-103).

Negre, A., Bensignor, E. and Guillot, J. (2009) 'Evidence-based veterinary dermatology: a systematic review of interventions for Malassezia dermatitis in dogs', *Veterinary Dermatology,* **20**, (1), 1-12.

Noli, C. and Auxilia, S. T. (2005) 'Treatment of canine Old World visceral leishmaniasis: a systematic review', *Veterinary Dermatology,* **16**, (4), 213-232.

Nuttall, T. and Cole, L. K. (2007) 'Evidence-based veterinary dermatology: a systematic review of interventions for treatment of Pseudomonas otitis in dogs', *Veterinary Dermatology,* **18**, (2), 69-77.

O'Connor, A. M., Wellman, N. G., Evans, R. B. and Roth, D. R. (2006) 'A review of randomized clinical trials reporting antibiotic treatment of infectious bovine keratoconjunctivitis in cattle', *Animal Health Research Reviews,* **7**, (1-2), 119-127.

Olivry, T. and Bizikova, P. (2010) 'A systematic review of the evidence of reduced allergenicity and clinical benefit of food hydrolysates in dogs with cutaneous adverse food reactions', *Veterinary Dermatology,* **21**, (1), 32-41.

Olivry, T., Foster, A. P., Mueller, R. S., McEwan, N. A., Chesney, C. and Williams, H. C. (2010) 'Interventions for atopic dermatitis in dogs: A systematic review of randomized controlled trials', *Veterinary Dermatology,* **21**, (1), 4-22.

Parr, M. J., Bouillon, B., Brohi, K., Dutton, R. P., Hauser, C. J., Hess, J. R., Holcomb, J. B., Kluger, Y., MacKway-Jones, K., Rizoli, S. B., Yukioka, T. and Hoyt, D. B. (2008) 'Traumatic coagulopathy: Where are the good experimental models?', *Journal of Trauma Injury, Infection and Critical Care,* **65**, (4), 766-771.

Perlroth, J., Kuo, M., Tan, J., Bayer, A. S. and Miller, L. G. (2008) 'Adjunctive use of rifampin for the treatment of Staphylococcus aureus infections: A systematic review of the literature', *Archives of Internal Medicine,* **168**, (8), 805-819.

Ploumis, A., Yadlapalli, N., Fehlings, M. G., Kwon, B. K. and Vaccaro, A. R. (2010) 'A systematic review of the evidence supporting a role for vasopressor support in acute SCI', *Spinal Cord,* **48**, (5), 356-362.

Ravnskov, U. (2005) 'Experimental glomerulonephritis induced by hydrocarbon exposure: A systematic review', *BMC Nephrology,* **6**, (15).

Renvert, S., Polyzois, I. and Maguire, R. (2009) 'Re-osseointegration on previously contaminated surfaces: a systematic review', *Clinical Oral Implants Research,* **20 Suppl 4**, 216-227.

Reynolds, J. C., Rittenberger, J. C. and Menegazzi, J. J. (2007) 'Drug administration in animal studies of cardiac arrest does not reflect human clinical experience', *Resuscitation,* **74**, (1), 13-26.

Sammour, T., Mittal, A., Loveday, B. P. T., Kahokehr, A., Phillips, A. R. J., Windsor, J. A. and Hill, A. G. (2009) 'Systematic review of oxidative stress associated with pneumoperitoneum', *British Journal of Surgery,* **96**, (8), 836-850.

Sargeant, J. M., Amezcua, M. R., Rajic, A. and Waddell, L. (2007) 'Pre-harvest interventions to reduce the shedding of E. coli O157 in the faeces of weaned domestic ruminants: a systematic review', *Zoonoses and Public Health,* **54**, (6-7), 260-277.

Schachtrupp, A., Wauters, J. and Wilmer, A. (2007) 'What is the best animal model for ACS?', *Acta Clinical Belgica Supplement,* (1), 225-232.

Sculean, A., Nikolidakis, D. and Schwarz, F. (2008) 'Regeneration of periodontal tissues: combinations of barrier membranes and grafting materials - biological foundation and preclinical evidence: a systematic review', *Journal of Clinical Periodontology,* **35**, (8 Suppl), 106-116.

Sena, E., Wheble, P., Sandercock, P. and Macleod, M. (2007) 'Systematic review and meta-analysis of the efficacy of tirilazad in experimental stroke', *Stroke,* **38**, (2), 388-394.

Sheng, S. R., Wang, X. Y., Xu, H. Z., Zhu, G. Q. and Zhou, Y. F. (2010) 'Anatomy of large animal spines and its comparison to the human spine: A systematic review', *European Spine Journal,* **19**, (1), 46-56.

Silverlas, C., Bjorkman, C. and Egenvall, A. (2009) 'Systematic review and meta-analyses of the effects of halofuginone against calf cryptosporidiosis', *Preventive Veterinary Medicine,* **91**, (2-4), 73-84.

Sniekers, Y. H., Weinans, H., Bierma-Zeinstra, S. M., van Leeuwen, J. P. T. M. and van Osch, G. J. V. M. (2008) 'Animal models for osteoarthritis: the effect of ovariectomy and estrogen treatment - a systematic approach', *Osteoarthritis and Cartilage,* **16**, (5), 533-541.

Steffan, J., Favrot, C. and Mueller, R. (2006) 'A systematic review and meta-analysis of the efficacy and safety of cyclosporin for the treatment of atopic dermatitis in dogs', *Veterinary Dermatology,* **17**, (1), 3-16.

Strom, J. O., Theodorsson, A. and Theodorsson, E. (2009) 'Dose-related neuroprotective versus neurodamaging effects of estrogens in rat cerebral ischemia: A systematic analysis', *Journal of Cerebral Blood Flow and Metabolism,* **29**, (8), 1359-1372.

Tapuria, N., Kumar, Y., Habib, M. M., Abu Amara, M., Seifalian, A. M. and Davidson, B. R. (2008) 'Remote ischemic preconditioning: a novel protective method from ischemia reperfusion injury--a review', *Journal of Surgical Research,* **150**, (2), 304-330.

Trevitt, C. R. and Collinge, J. (2006) 'A systematic review of prion therapeutics in experimental models', *Brain,* **129**, (9), 2241-2265.

Trksak, G. H., Glatt, S. J., Mortazavi, F. and Jackson, D. (2007) 'A meta-analysis of animal studies on disruption of spatial navigation by prenatal cocaine exposure', *Neurotoxicology and Teratology,* **29**, (5), 570-577.

Van Der Worp, H. B., Sena, E. S., Donnan, G. A., Howells, D. W. and Macleod, M. R. (2007) 'Hypothermia in animal models of acute ischaemic stroke: A systematic review and meta-analysis', *Brain,* **130**, (12), 3063-3074.

Vlastarakos, P. V., Nikolopoulos, T. P., Tavoulari, E., Papacharalambous, G. and Korres, S. (2008) 'Auditory neuropathy: Endocochlear lesion or temporal processing impairment? Implications for diagnosis and management', *International Journal of Pediatric Otorhinolaryngology,* **72**, (8), 1135-1150.

Vlastarakos, P. V., Nikolopoulos, T. P., Tavoulari, E., Papacharalambous, G., Tzagaroulakis, A. and Dazert, S. (2008) 'Sensory cell regeneration and stem cells: what we have already achieved in the management of deafness', *Otology and Neurotology,* **29**, (6), 758-768.

von Bohl, M. and Kuijpers-Jagtman, A. M. (2009) 'Hyalinization during orthodontic tooth movement: a systematic review on tissue reactions', *European Journal of Orthodontics,* **31**, (1), 30-36.

Warner, D. and Brietzke, S. E. (2008) 'Mitomycin C and airway surgery: How well does it work?', *Otolaryngology Head and Neck Surgery*, **138**, (6), 700-709.

Weigl, M., Tenze, G., Steinlechner, B., Skhirtladze, K., Reining, G., Bernardo, M., Pedicelli, E. and Dworschak, M. (2005) 'A systematic review of currently available pharmacological neuroprotective agents as a sole intervention before anticipated or induced cardiac arrest', *Resuscitation*, **65**, (1), 21-39.

Wellman, N. G. and O'Connor, A. M. (2007) 'Meta-analysis of treatment of cattle with bovine respiratory disease with tulathromycin', *Journal of Veterinary Pharmacology and Therapeutics*, **30**, (3), 234-241.

Wheble, P. C. R., Sena, E. S. and Macleod, M. R. (2008) 'A systematic review and meta-analysis of the efficacy of piracetam and piracetam-like compounds in experimental stroke', *Cerebrovascular Diseases*, **25**, (1-2), 5-11.

Wiegand, A. and Attin, T. (2008) 'Efficacy of enamel matrix derivatives (Emdogain) in treatment of replanted teeth--a systematic review based on animal studies', *Dental Traumatology*, **24**, (5), 498-502.

Willmot, M., Gibson, C., Gray, L., Murphy, S. and Bath, P. (2005) 'Nitric oxide synthase inhibitors in experimental ischemic stroke and their effects on infarct size and cerebral blood flow: A systematic review', *Free Radical Biology and Medicine*, **39**, (3), 412-425.

Wira, C. R., Becker, J. U., Martin, G. and Donnino, M. W. (2008) 'Anti-arrhythmic and vasopressor medications for the treatment of ventricular fibrillation in severe hypothermia: A systematic review of the literature', *Resuscitation*, **78**, (1), 21-29.

Zhang, J., Xie, X., Li, C. and Fu, P. (2009) 'Systematic review of the renal protective effect of Astragalus membranaceus (root) on diabetic nephropathy in animal models', *Journal of Ethnopharmacology*, **126**, (2), 189-196.

Appendix D: Assessment form

Scoring form Systematic Reviews

No.	Article	Search Strategy								
		MDB	DBN	OSE	D	Y	SSR	SSQ	S	LR
1										
2										
3										
4										
...										
...										
77										
78										
79										
80										

MDB: Multiple Databases used
DBN: Database name
OSE: Other sources of evidence
D: Search date
Y: Years included
SSR: Search strategy repeatability
SSQ: Technical quality of search strategy
S: Summary of search strategy

LR: Language restrictions

Appendix E: Step-by-step guide for effectively finding all relevant animal studies

Step			Action	Method
1			Define research question	PICOT mnemonic
2			Define sources to search in	At least 2 bibliographic databases Other sources of evidence
3			Develop database specific search strings	
	P		PubMed string	
		1	Animal Studies Set	Use filter by Hooijmans *et al.* (2010b)
		2	Create a Disease Set	Define all relevant synonyms, singular/plural and spellings Select and compose MeSH terms [MeSH] Select keywords for title and abstract [tiab] Connect with Boolean Operators
		3	Create an Intervention set	Define all relevant synonyms, singular/plural and spellings Select and compose MeSH terms [MeSH] Select keywords for title and abstract [tiab] Connect with Boolean Operators
		4	Combine Sets	'Animal Studies Set' AND 'Disease Set' AND 'Intervention Set'
	E		EMBASE string	
		1	Animal Studies Set	de Vries *et al.* (in preparation)
		2	Create a Disease Set	Define all relevant synonyms, singular/plural and spellings Select and compose Emtree terms exp <term> / Select keywords for title and abstract .ti,ab. Connect with Boolean Operators
		3	Create an Intervention set	Define all relevant synonyms, singular/plural and spellings Select and compose Emtree terms exp <term> / Select keywords for title and abstract .ti,ab. Connect with Boolean Operators
		4	Combine Sets	'Animal Studies Set' AND 'Disease Set' AND 'Intervention Set'
4			Execute search strings	
5			Cross reference results in order to remove double citations	

www.ingramcontent.com/pod-product-compliance
Lightning Source LLC
Chambersburg PA
CBHW070332190526
45169CB00005B/1860